CONTROL NETWORKS FOR THE GALILEAN SATELLITES: NOVEMBER 1979

PREPARED FOR THE JET PROPULSION LABORATORY
AND THE NATIONAL AERONAUTICS AND SPACE ADMINISTRATION

MERTON E. DAVIES, THOMAS A. HAUGE,
FRANK Y. KATAYAMA, JAMES A. ROTH

R-2532-JPL/NASA NOVEMBER 1979

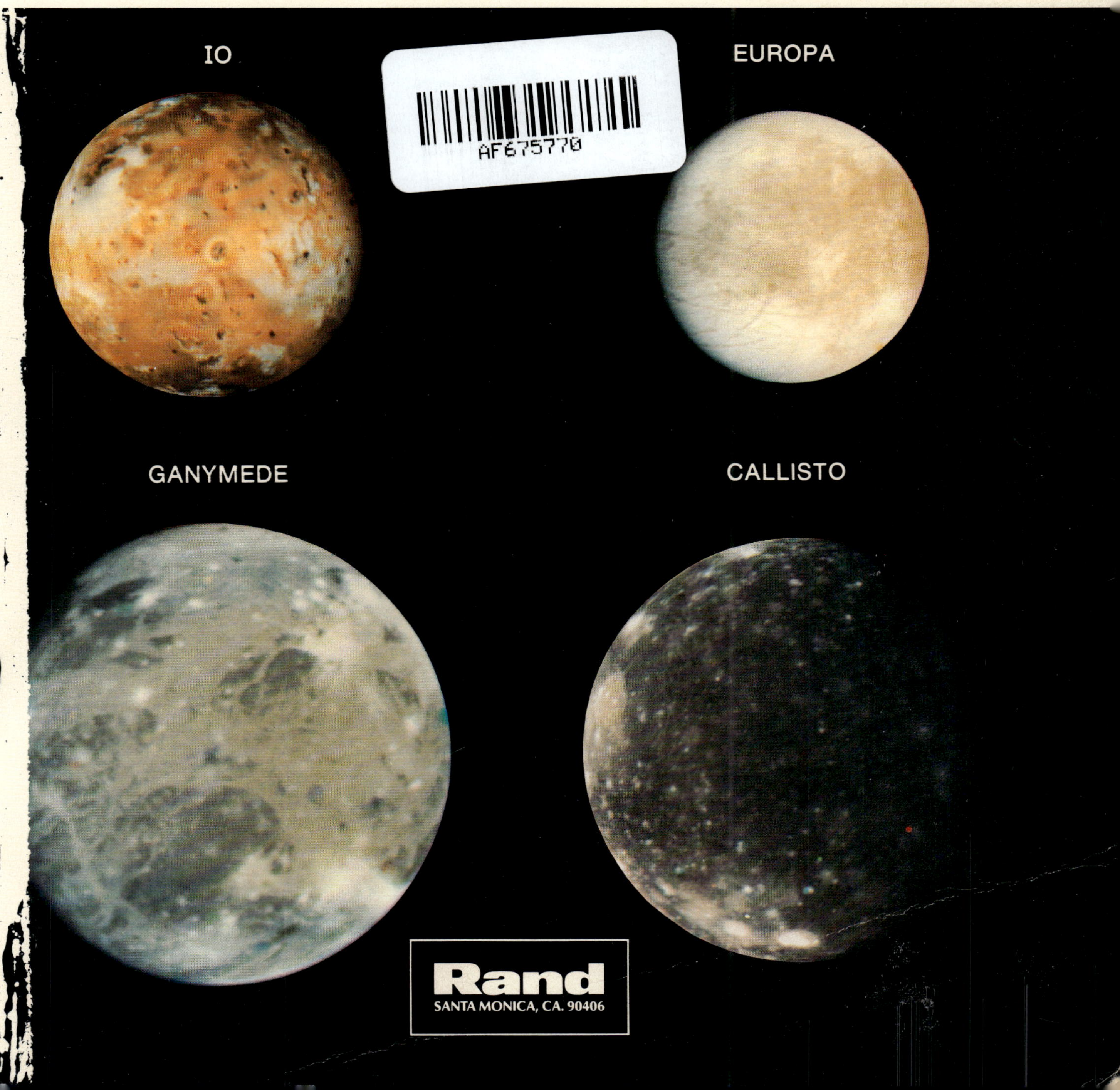

This research is sponsored by the Jet Propulsion Laboratory under Contract No. 953613 and the National Aeronautics and Space Administration under Contract No. NASW-3210.

Library of Congress Cataloging in Publication Data

Main entry under title:

Control networks for the Galilean satellites,
November 1979.

([Report] - Rand Corporation ; R-2532-
JPL/NASA-
Bibliography: p.
1. Space photography--Technique.
2. Satellites--Jupiter--Photographs from space.
I. Davies, Merton E. II. California.
Institute of Technology, Pasadena. Jet Propulsion
Laboratory. III. United States. National
Aeronautics and Space Administration.
IV. Series: Rand Corporation. Rand report ;
R-2532-JPL/NASA-
AS36.R3 R-2532, etc. [TR713] 081s [522'.63]
ISBN 0-8330-0197-3 (v. 1) 79-27960

Published by The Rand Corporation

CONTROL NETWORKS FOR THE GALILEAN SATELLITES: NOVEMBER 1979

PREPARED FOR THE JET PROPULSION LABORATORY
AND THE NATIONAL AERONAUTICS AND SPACE ADMINISTRATION

MERTON E. DAVIES, THOMAS A. HAUGE,
FRANK Y. KATAYAMA, JAMES A. ROTH

R-2532-JPL/NASA
NOVEMBER 1979

Rand
SANTA MONICA, CA. 90406

PREFACE

The Imaging Science Team of the Voyager mission was established by the National Aeronautics and Space Administration in 1972 with Bradford A. Smith as team leader. Over the years, as new members were added, the team grew from the original 9 scientists to the 22 who were present at the time of the Jupiter encounters in March and July 1979. Those encounters yielded many surprises and discoveries and, of course, spectacular pictures. Many follow-up studies are now under way by members of the team.

This report presents the first results on the development of the coordinate systems and control networks of the Galilean satellites. The coordinates of features, the pictures, and the camera orientation matrices are used in preparing surface maps, computer mosaics, and spectral maps. The mean radii of the satellites as measured by the control net are more accurate than those obtained by fitting circles to the satellite limbs on individual pictures.

This report presents results of research carried out under JPL contract 953613 and NASA contract NASW-3210.

SUMMARY

Pictures of the four Galilean satellites were taken as the two Voyager spacecraft approached Jupiter during March and July 1979, and nearly full coverage of all the satellites was obtained by each flyby as the satellites rotated in front of the cameras. High-resolution mosaics of the satellites were obtained during the periods of closest encounter.

The faceplates of the vidicon cameras have reseaux of 202 points to permit removal of geometric distortions. Pictures of stars in the Pleiades cluster were used to measure the focal lengths of the optical systems and to calibrate the cameras geometrically.

Control nets of the Galilean satellites have been computed photogrammetrically and mean radii measured. This work is continuing, and improved results will emerge as more data are incorporated in the computations. At this time the mean radii of the satellites are: Io, 1816 ± 5 km; Europa, 1563 ± 10 km; Ganymede, 2638 ± 10 km; and Callisto, 2410 ± 10 km. The control nets contain: Io, 307 points, 159 pictures; Europa, 86 points, 46 pictures; Ganymede, 227 points, 71 pictures; and Callisto, 291 points, 95 pictures. The pictures in the control nets are identified, the coordinates of the control points are given, and identifications of some of the control points are shown on figures.

ACKNOWLEDGMENTS

The authors are indebted to their colleagues on the Imaging Science Team and the many hundreds of scientists and engineers at the Jet Propulsion Laboratory (JPL) who made the encounters so successful. In particular, we are grateful for the efforts of Mary Brownell and Candice Hansen of JPL in preparing the picture-taking sequences for this experiment. We would like to thank Peter Kupferman, Larry Tietze, and Linda Morabito of JPL for star exposure data, star coordinates, and star plots. We are indebted to Leonard Dicken, Andrey Sergeyevsky, and James Campbell of the Voyager Navigation Team for trajectory updates and to Frances Popescu of JPL for putting these data in machine-readable form for the Rand computer.

The maps used in the figures were prepared by the USGS, Flagstaff, under the direction of Raymond M. Batson. Patricia M. Bridges (Io and Callisto) and Jay L. Inge (Europa and Ganymede) were responsible for the surface interpretations and beautiful airbrush renditions.

The authors would like to thank Stephen H. Dole and Louis N. Rowell of Rand for careful reviews of the manuscript.

CONTENTS

FIGURES

TABLES

I. INTRODUCTION

On March 5, 1979, Voyager 1 flew by Jupiter and started on its path to Saturn. During its encounter with Jupiter, about 18,000 pictures of the planetary system were taken (Smith et al., 1979a). Voyager 2 encountered Jupiter on July 9, 1979, and recorded about 15,000 pictures (Smith et al., 1979b). Most of the pictures were of the planet; less than 10 percent were of the four Galilean satellites. The aiming points of both encounters were chosen to send the spacecraft on to Saturn and to maximize the number of close encounters with the Galilean satellites. The success of this strategy can be seen in Table 1, which gives the closest approaches to each satellite.

Table 1

SELECTED SATELLITE ORBITAL AND SPACECRAFT TRAJECTORY PARAMETERS

Satellite	Mean Distance from Jupiter (km)	Mean Orbital Period (days)	Closest Approach by	
			Voyager 1 (km)	Voyager 2 (km)
Io (J1)	421,600	1.769	20,570	1,129,900
Europa (J2)	670,900	3.551	733,760	205,720
Ganymede (J3)	1,070,000	7.155	114,710	62,130
Callisto (J4)	1,880,000	16.689	126,400	214,930

Each spacecraft carried two slow-scan vidicon cameras, one with 1500-mm focal length optics and the other with 200-mm focal length optics (Smith et al., 1977). The raster is 800 × 800 pixels (picture elements) with 8-bit encoding. For reference in assessing the image quality at the approach distances given in Table 1, the 1500-mm camera gives a 1-km, 2-pixel surface resolution from a distance of 50,000 km.

The Galilean satellites are in synchronous rotation, so the same region always faces Jupiter in the same way that the same side of the Moon always faces the Earth. Thus their rotation periods are the same as their orbital periods (see Table 1). The picture-taking sequences were designed to take a series of pictures of each satellite through a set of color filters approximately every 15° of satellite longitude as the spacecraft approached the Jupiter system. Each spacecraft thus obtained essentially full-coverage pictures of each satellite. As the spacecraft got closer to each satellite, it became necessary to mosaic to obtain full coverage. At closest approach, pictures were taken with the wide-angle-lens camera because smear sometimes limited the resolution obtainable with the narrow-angle-lens camera. Figure 1, which shows the resolution versus satellite longitude coverage for the Voyager 1 and 2 encounters, reflects this sequence strategy (Smith et al., 1979b).

The computation of geodetic control nets of the Galilean satellites and their sizes and shapes was one of the scientific objectives of the Voyager mission (Smith et al., 1977). The computational methods are essentially the same as those employed at Mars (Davies, 1972; Davies and Arthur, 1973) and at Mercury (Davies and Batson, 1975). Results from the analytical triangulation (the control net computation) are required for positional data in the preparation of maps. Auxiliary data from the computation are important for the registration of color images and computer mosaics.

This report will discuss (1) the use of star field pictures to compute the focal lengths of the cameras and the geometric relationship between the narrow- and wide-angle cameras, (2) the description of the coordinate systems of the Galilean satellites, and (3) the status of the control net computations. Coordinates of the control points and illustrations of some of their locations are given.

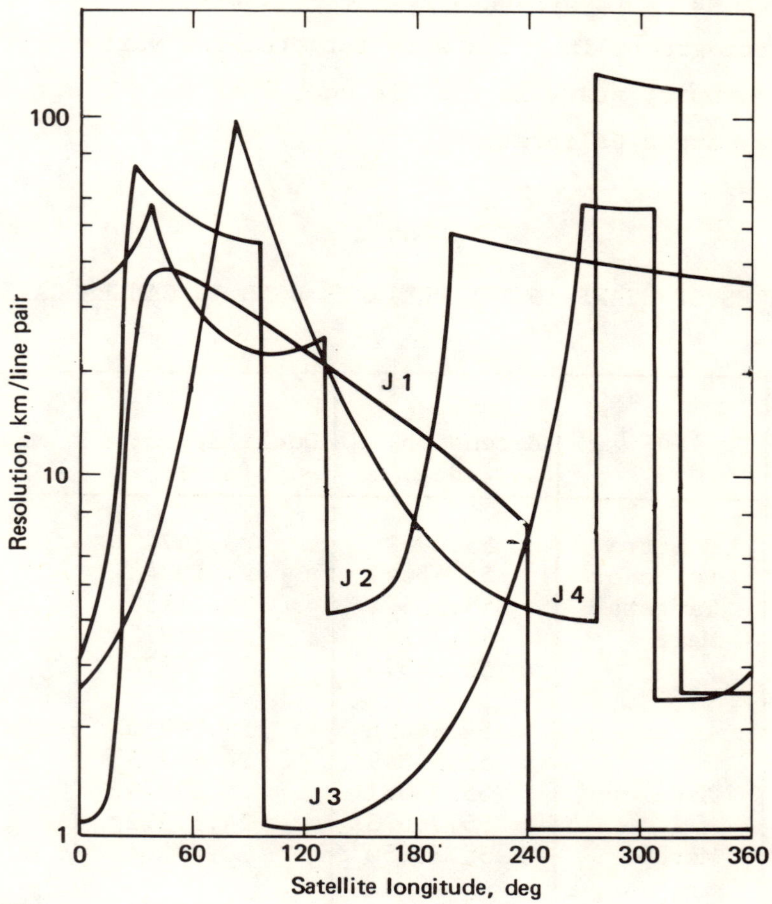

NOTE:
Each of the Galilean satellites (Io, J1; Europa, J2; Ganymede, J3; and Callisto, J4) was photographed throughout the final orbit prior to each Voyager's closest approach. Sharp discontinuities in resolution correspond to the terminator longitude at the times of closest approach.

Fig. 1—Combined resolution versus satellite longitude coverage for the Voyager encounters

II. STAR CALIBRATION

Pictures of star groups were taken during the spacecraft's cruise phase for refinement of the pointing capability of the scan platform and for geometric calibration of the vidicon cameras. The Pleiades open cluster was a favorite target for calibration frames, whereas selected star groups in Orion were targeted for variety. Table 2 lists the coordinates of stars in the Pleiades that were recorded on pictures used in the camera calibration.

Table 2

COORDINATES OF STARS IN THE PLEIADES USED IN CAMERA CALIBRATIONS

Number	Star Name	Right Ascension, α (deg)	Declination, δ (deg)	Magnitude
1	Electra	55.47539	23.95740	3.81
2	Celaeno	55.45647	24.13343	5.43
3	Taygeta	55.55671	24.31159	4.37
4	Maia	55.71166	24.21272	4.02
5		56.34570	23.26874	5.51
6		54.44498	25.16930	6.15
7		55.54349	24.68355	5.63
8	Asterope	55.73081	24.39970	5.85
9		55.76615	24.37312	6.46
10	Merope	55.83850	23.79384	4.25
11	Alcyone	56.12694	23.95177	2.96
12	Atlas	56.54611	23.90180	3.80
13	Pleione	56.55175	23.98507	5.18
14		56.68839	23.56071	6.11
15		56.82630	25.42926	5.38

The photosensitive surface of each vidicon contains a reseau of 202 points; each point is a square about 45 μm (3 pixels) on a side. The origin of the image coordinate system is the central reseau point, and the horizontal axis coincides with the horizontal line of reseau points through the central reseau. Before assembly, the coordinates of each point were measured in mm units (Benesh and Jepsen, 1978).

A typical star picture taken in the calibration sequence is shown in Fig. 2. An overlay grid is programmed for the computer to aid in counting pixels. The star locations are measured by counting pixel coordinates of the star and three surrounding reseau points, then the star image coordinates are determined by interpolation between the reseau points. From the coordinates of the stars (Table 2), the focal length of the camera and the three angles of the camera orientation matrix (C matrix) can be determined by the method of least squares. Results from focal length measurements are summarized in Table 3.

The Voyager spacecraft are able to shutter both cameras simultaneously. All of the frames listed in Table 3 were simultaneous exposures; in this mode the narrow-angle camera is read out first--thus frames 17373.05 and 17373.06 are a pair, frames 17373.09 and 17373.10 are a pair, and so forth. As mentioned above, each time there is a solution for the camera focal length, there is also a solution for the camera orientation matrix, C. Thus, by matrix multiplication, the matrix $C_{NA}C_{WA}^{-1}$, which relates the orientation matrix of the wide-angle camera to that of the narrow-angle camera, can be determined from the simultaneous exposures. Mean values of this matrix are given in Table 4.

Fig. 2—Voyager 2 wide-angle picture of the Pleiades with computer overlaid grid to aid counting pixel coordinates of stars and reseau points (frame FDS 10453.33)

Table 3

FOCAL LENGTH MEASUREMENTS FROM STAR PICTURES

Frame Number	Calibration Stars from Table 2	Number of Stars	Focal Length (mm)
Voyager 1			
17373.06	1,4,6,7,13,15	6	200.594
17373.10	2,5,6,7,13,15	6	200.143
17373.18	1,5,7,12,15	5	200.241
17373.22	4,5,7,13,15	5	200.199
17373.31	2,5,7,13,15	5	200.287
			200.293 ± 0.177 average
17373.05	1,2,4	3	1500.030
17373.09	1,2,3,4	4	1500.368
17373.17	1,2,4	3	1500.262
17373.21	1,2,4	3	1500.522
17373.30	3,4,8	3	1499.746
			1500.19 ± 0.30 average
Voyager 2			
10453.33	1,2,3,4,5,6,7,8, 9,10,11,12,13,14	14	200.938
10453.41	1,2,3,4,5,6,7,8, 9,10,11,12,13,14	14	200.827
10453.45	1,2,3,4,5,7,8, 9,10,11,12,13,14	13	200.883
10456.11	Orion	5	200.433
			200.770 ± 0.229 average
10453.32	1,2,3,4	4	1503.621
10453.40	1,2,4	3	1503.392
10453.44	1,2,3,4	4	1503.006
10456.10	Orion	3	1503.935
			1503.49 ± 0.39 average

Table 4

THE MATRIX ($C_{NA}C_{WA}^{-1}$) RELATING THE CAMERA AIMING AND ROTATION DIRECTIONS OF THE TWO CAMERAS ON EACH SPACECRAFT

$C_{NA}C_{WA}^{-1}$

Voyager 1

$$\begin{bmatrix} 0.9999950588 & -0.0031011413 & 0.0005151430 \\ 0.0031009104 & 0.9999950916 & 0.0004483927 \\ -0.0005165310 & -0.0004467931 & 0.9999997668 \end{bmatrix}$$

Voyager 2

$$\begin{bmatrix} 0.9999966464 & -0.0025304697 & -0.0005512367 \\ 0.0025303754 & 0.9999967838 & -0.0001717796 \\ 0.0005516696 & 0.0001703842 & 0.9999998333 \end{bmatrix}$$

III. THE SATELLITE COORDINATE SYSTEMS

Since the Galilean satellites are in synchronous rotation, their axes of rotation should be approximately normal to their orbital planes (Peale, 1977). The Voyager pictures appear to bear out this deduction. As defined at the 1973 International Astronomical Union (IAU) General Assembly, the prime meridian passes through the subplanetary intersection of the satellite's equator and the plane containing the centers of the satellite, Jupiter, and the Sun at the time of the first superior heliocentric conjunction of the satellite and the planet after 1950.0.

The direction of the north pole of the satellite is specified by its right ascension, α_o, and declination, δ_o. Its prime meridian is specified by the angle W that is measured along the satellite's equator in an easterly direction from the ascending node Q of the satellite's equator on the standard Earth equator to the point B where the prime meridian crosses the satellite's equator (see Fig. 3). W varies linearly with time due to the uniform rotation of the satellite. In addition,

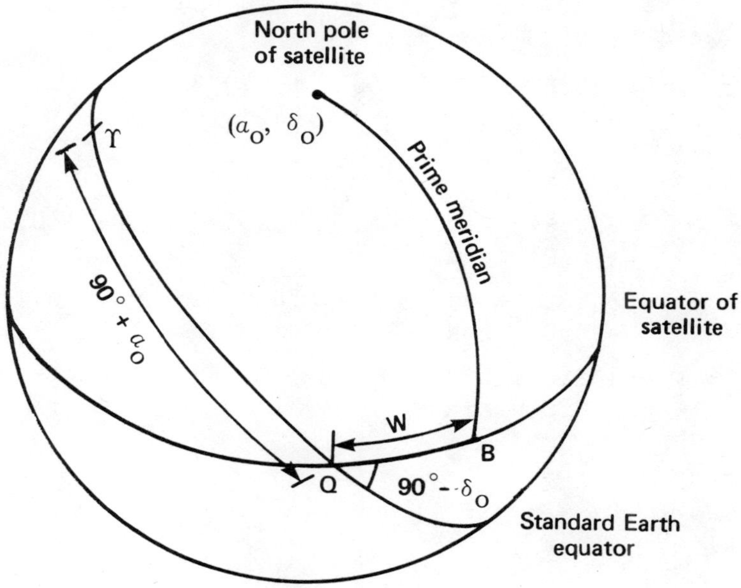

Fig. 3—Reference system used to define orientation of the satellite

α_o, δ_o, and W vary with time due to the precession of the axis of rotation of the satellite. The equations for α_o, δ_o, and W for the Galilean satellites have been derived by Lieske (1979) and adopted by the IAU (*Transactions 1979*). The expressions are

Io

$$\alpha_o = 268^{\circ}.002 - 0^{\circ}.0085T + 0^{\circ}.094 \sin \text{♃}_1 + 0^{\circ}.024 \sin \text{♃}_2$$

$$\delta_o = 64^{\circ}.504 + 0^{\circ}.0033T + 0^{\circ}.040 \cos \text{♃}_1 + 0^{\circ}.011 \cos \text{♃}_2$$

$$W = 262^{\circ}.7 + 203^{\circ}.4889538d - 0^{\circ}.085 \sin \text{♃}_1 - 0^{\circ}.022 \sin \text{♃}_2$$

Europa

$$\alpha_o = 268^{\circ}.029 - 0^{\circ}.0085T + 1^{\circ}.086 \sin \text{♃}_2 + 0^{\circ}.060 \sin \text{♃}_3 + 0^{\circ}.015 \sin \text{♃}_4 + 0^{\circ}.009 \sin \text{♃}_5$$

$$\delta_o = 64^{\circ}.516 + 0^{\circ}.0033T + 0^{\circ}.468 \cos \text{♃}_2 + 0^{\circ}.026 \cos \text{♃}_3 + 0^{\circ}.007 \cos \text{♃}_4 + 0^{\circ}.002 \cos \text{♃}_5$$

$$W = 156^{\circ}.9 + 101^{\circ}.3747235d - 0.980 \sin \text{♃}_2 - 0^{\circ}.054 \sin \text{♃}_3 - 0^{\circ}.014 \sin \text{♃}_4 - 0^{\circ}.008 \sin \text{♃}_5$$

Ganymede

$$\alpha_o = 268^{\circ}.149 - 0^{\circ}.0085T - 0^{\circ}.037 \sin \text{♃}_2 + 0^{\circ}.431 \sin \text{♃}_3 + 0^{\circ}.091 \sin \text{♃}_4$$

$$\delta_o = 64^{\circ}.574 + 0^{\circ}.0033T - 0^{\circ}.016 \cos \text{♃}_2 + 0^{\circ}.186 \cos \text{♃}_3 + 0^{\circ}.039 \cos \text{♃}_4$$

$$W = 195^{\circ}.8 + 50^{\circ}.3176081d + 0^{\circ}.033 \sin \text{♃}_2 - 0^{\circ}.389 \sin \text{♃}_3 - 0^{\circ}.082 \sin \text{♃}_4$$

Callisto

$$\alpha_o = 268^{\circ}.678 - 0^{\circ}.0085T - 0^{\circ}.068 \sin \text{♃}_3 + 0^{\circ}.590 \sin \text{♃}_4 + 0^{\circ}.010 \sin \text{♃}_6$$

$$\delta_o = 64^{\circ}.830 + 0^{\circ}.0033T - 0^{\circ}.029 \cos \text{♃}_3 + 0^{\circ}.254 \cos \text{♃}_4 - 0.004 \cos \text{♃}_6$$

$$W = 158^{\circ}.0 + 21^{\circ}.5710715d + 0^{\circ}.061 \sin \text{♃}_3 - 0^{\circ}.533 \sin \text{♃}_4 - 0^{\circ}.009 \sin \text{♃}_6$$

where

$$\text{♃}_1 = 19^{\circ}.2 + 4850^{\circ}.7T$$

$$\text{♃}_2 = 120^{\circ}.8 + 1191^{\circ}.3T$$

$$\text{♃}_3 = 349^{\circ}.5 + 262^{\circ}.1T$$

$$\text{♃}_4 = 198^{\circ}.3 + 64^{\circ}.3T$$

$$\text{♃}_5 = 241^{\circ}.6 + 2382^{\circ}.6T$$

$$\text{♃}_6 = 317^{\circ}.7 + 6070^{\circ}.0T$$

NOTES: α_o, δ_o are standard equatorial coordinates of 1950.0.

T is the interval in Julian ephemeris centuries (of 36525 days) from the standard epoch.

d is the interval in ephemeris days from the standard epoch.

The standard epoch is 1950 January 1.0 ET, i.e., JED2433282.5.

A point, P, on the surface of a satellite has cartographic coordinates latitude ϕ, west longitude λ, and radius R, and rectangular coordinates X, Y, Z, where $X = R \cos\phi \cos(360^\circ - \lambda)$, $Y = R \cos\phi \sin(360^\circ - \lambda)$, and $Z = R \sin\phi$. Because the X, Y, Z coordinate system is right-handed, $360^\circ - \lambda$ is used in the expressions for X and Y. The Z axis is the axis of rotation of the satellite with north positive. The X axis lies in the plane of the equator, positive in the direction of 0° longitude. The Y axis lies in the plane of the equator, positive in the direction of 270° west longitude. The standard equatorial coordinates of 1950.0 of the point P_x, P_y, P_z can be expressed as

$$\begin{bmatrix} P_x \\ P_y \\ P_z \end{bmatrix} = MV \begin{bmatrix} X \\ Y \\ Z \end{bmatrix}$$

where

$$M^T = \begin{bmatrix} 1 & 0 & 0 \\ 0 & \cos(90^\circ - \delta_o) & \sin(90^\circ - \delta_o) \\ 0 & -\sin(90^\circ - \delta_o) & \cos(90^\circ - \delta_o) \end{bmatrix} \begin{bmatrix} \cos(\alpha_o + 90^\circ) & \sin(\alpha_o + 90^\circ) & 0 \\ -\sin(\alpha_o + 90^\circ) & \cos(\alpha_o + 90^\circ) & 0 \\ 0 & 0 & 1 \end{bmatrix}$$

and

$$V = \begin{bmatrix} \cos W & -\sin W & 0 \\ \sin W & \cos W & 0 \\ 0 & 0 & 1 \end{bmatrix}.$$

If a picture containing P is taken by the spacecraft at S_x, S_y, S_z, the coordinates X_c, Y_c of P on the picture are given by

$$X_c = \frac{\xi}{\zeta} f, \quad Y_c = \frac{\eta}{\zeta} f,$$

where $$\begin{bmatrix} \xi \\ \eta \\ \zeta \end{bmatrix} = C \begin{bmatrix} P_x \\ P_y \\ P_z \end{bmatrix} - C \begin{bmatrix} S_x \\ S_y \\ S_z \end{bmatrix},$$

and f is the calibrated principal distance (focal length) and C is the transformation matrix from standard coordinates of 1950.0 into the camera coordinate system. X_c, Y_c, f are expressed in millimeters and R, P_x, P_y, P_z, S_x, S_y, S_z are in kilometers.

Coordinates of the point P are measured on the picture by counting pixels and then removing the vidicon distortions and scaling the pixel coordinates to millimeter coordinates X_o, Y_o at the faceplate of the vidicon. The reseau is used in this transformation. The pixel measurements on the pictures are estimated to the one-tenth pixel and in general are repeatable to a few tenths of a pixel.

Standard photogrammetric methods are used to solve for the unknowns (for instance, see Davies and Arthur, 1973). Approximate values of all parameters are required to initiate the analytical triangulation. The triangulation is a problem in least squares designed to minimize the sum of the squares of the residuals, i.e., $(X_o - X_c)$, $(Y_o - Y_c)$. Observation equations are expressed in terms of those parameters whose values are permitted to vary; the normal equations are formed and solved to give improved values to the desired parameters. In practice, the spacecraft positions S_x, S_y, S_z are never permitted to vary, and the angles of the C matrix are always variable, as are the latitude ϕ and longitude λ of the control points. The radius at the control points can be fixed, a single mean radius determined for all points, or the radius at each point determined independently. In the last case, to solve for the satellite radius at each control point requires excellent stereo pictures because the radii are highly correlated with the camera orientation angles (C matrix).

Closest approach to Jupiter took place 5 March 1979 with the Voyager 1 encounter and 9 July 1979 with the Voyager 2 encounter. The control net computations for each satellite incorporate pictures from both encounters. Although the direction of the north pole of the satellites does vary with time, the time between the two encounters is not significant and the M matrix is treated as a constant. For Io all of the near-encounter pictures were acquired by Voyager 1; α_o, δ_o were evaluated at 5 March 1979 (JED 2443937.5). For Europa, all of the near-encounter pictures were acquired by Voyager 2; α_o, δ_o were evaluated at 9 July 1979 (JED 2444063.5). Near-encounter pictures of Ganymede and Callisto were taken by both Voyager 1 and 2, so α_o, δ_o were evaluated at 7 May 1979 (JED 2444000.5). The M matrices of the satellites are given in Table 5.

Table 5

M MATRICES FOR THE GALILEAN SATELLITES

$$\text{Io} \quad \begin{bmatrix} 0.99939690 & 0.03135277 & -0.01492764 \\ -0.03472507 & 0.90234109 & -0.42962161 \\ 0.0 & 0.42988087 & 0.90288562 \end{bmatrix}$$

$$\text{Europa} \quad \begin{bmatrix} 0.99986827 & 0.01463016 & -0.00702844 \\ -0.01623085 & 0.90126082 & -0.43297287 \\ 0.0 & 0.43302991 & 0.90137955 \end{bmatrix}$$

$$\text{Ganymede} \quad \begin{bmatrix} 0.99963410 & 0.02443864 & -0.01159387 \\ -0.02704930 & 0.90315438 & -0.42846295 \\ 0.0 & 0.42861978 & 0.90348496 \end{bmatrix}$$

$$\text{Callisto} \quad \begin{bmatrix} 0.99953359 & 0.02759109 & -0.01308941 \\ -0.03053851 & 0.90306357 & -0.42841987 \\ 0.0 & 0.42861978 & 0.90348496 \end{bmatrix}$$

IV. THE CONTROL NETS OF THE GALILEAN SATELLITES

The control nets of the satellites are computed by means of single-block analytical triangulations. For convenience, the normal equations are solved by the conjugate gradient iterative method. The nets are updated frequently as additional points, pictures, and measurements are added to the data set. In updating the nets, the radii at the control points are held constant at the value of the mean radius and the point latitude and longitude vary as do the C matrices. Periodically, the mean radius is permitted to vary, thus leading to an improved measurement of the satellite radius. The status of the control net computations is summarized in Table 6.

Table 6

CURRENT STATUS OF CONTROL NETS OF THE GALILEAN SATELLITES

Parameter	Io	Europa	Ganymede	Callisto
Points	307	86	227	291
Pictures	159	46	71	95
Observation equations	4382	1058	2136	2366
Normal equations	1091	310	667	867
Overdetermination factors	4.02	3.41	3.20	2.73
Standard error of measurements, mm	0.02059	0.02310	0.03211	0.02098
Mean radius, km	1816	1563	2638	2410

Following each encounter trajectory, studies are made by the Voyager Navigation Team to improve the positional data of the spacecraft, satellites, and Jupiter. These improved trajectory solutions are incorporated in the control net computations by updating the S_x, S_y, S_z coordinates in the expressions for X_c, Y_c.

The accuracy of the control net can be improved if some constraints can be placed on at least a few of the C matrices. This is accomplished by taking simultaneous wide- and narrow-angle pictures at selected times when two satellites are in the wide-angle frame. A knowledge of the locations of the two satellites in the wide-angle frame permits computation of the camera orientation matrix C_{WA}. The camera orientation matrix of the narrow-angle frame C_{NA} can be determined as $C_{NA} = C_{NA}C_{WA}^{-1}C_{WA}$, where $C_{NA}C_{WA}^{-1}$ is given in Section I. Star pictures also can be used for this purpose. Narrow-angle frames which have C matrices constrained by simultaneous wide-angle frames are listed in Table 7.

Table 7

NARROW-ANGLE FRAMES WITH C MATRICES CONSTRAINED BY SIMULTANEOUSLY EXPOSED WIDE-ANGLE FRAMES

Satellite	Rand Number	FDS[a] Frame	Picture Number
Io	10004	16322.18	1373J1-003
Europa	20022	16323.18	1433J1-003
	20033	16357.07	1663J1-002
Ganymede	30026	16289.36	1211J1-004
	30035	16356.55	1651J1-002
Callisto	40049	16321.59	1354J1-003
	40051	16323.04	1419J1-003

[a]Flight Data System

The centers of particular crater rims are defined as the control points on the Moon, Mars, and Mercury; craters are convenient for control points as they are easily identifiable under different lighting and viewing geometries, and picture coordinates are readily measured. Thus, on Ganymede and Callisto, control points are commonly associated with specific craters. However, on Io and Europa, craters are scarce, so points are defined in various ways. On Io, the points are frequently the centroid of dark albedo spots and corners or intersections of linear markings. On Europa, the points are usually the intersections of the many conspicuous long linear features that cover the surface.

Table 8 is a list of the pictures in the Io control net. In addition to the Rand number is the FDS number and the picture number. The camera which took the picture is identified by the code: 1 = Voyager 2 wide angle, 2 = Voyager 2 narrow angle, 3 = Voyager 1 wide angle,

Table 8

IO: PICTURES IN THE CONTROL NET

Rand Number	FDS Number	Camera[a]	Picture Number	Rand Number	FDS Number	Camera[a]	Picture Number
10004	1632218	4	1373J1-003	10146	1637540	4	0977J1-001
10005	1632222	4	1377J1-003	10147	1637542	4	0979J1-001
10006	1634522	4	0957J1-002	10027	1637750	4	1107J1-001
10008	1634738	4	1093J1-002	10199	1637544	4	0981J1-001
10185	1635130	4	1326J1-002	10028	1637752	4	1109J1-001
10186	1635132	4	1328J1-002	10029	1637754	4	1111J1-001
10009	1635134	4	1330J1-002	10030	1637756	4	1113J1-001
10187	1635136	4	1332J1-002	10149	1638209	4	1366J1-001
10010	1635138	4	1334J1-002	10031	1638211	4	1368J1-001
10188	1635140	4	1336J1-002	10148	1638213	4	1370J1-001
10011	1635436	4	1512J1-002	10032	1638219	4	1376J1-001
10013	1635724	4	1680J1-002	10033	1638227	4	1384J1-001
10014	1636044	4	0080J1-001	10034	1638235	4	1392J1-001
10181	1636046	4	0082J1-001	10096	1638854	4	1771J1-001
10015	1636048	4	0084J1-001	10097	1638856	4	1773J1-001
10182	1636050	4	0086J1-001	10036	1638858	4	1775J1-001
10183	1636052	4	0088J1-001	10098	1638900	4	1777J1-001
10184	1636054	4	0090J1-001	10099	1638902	4	1779J1-001
10016	1636826	4	0542J1-001	10100	1638904	4	1781J1-001
10017	1636832	4	0548J1-001	10037	1638906	4	1783J1-001
10172	1636834	4	0550J1-001	10101	1638908	4	1785J1-001
10018	1636838	4	0554J1-001	10102	1638910	4	1787J1-001
10173	1636844	4	0560J1-001	10103	1638912	4	1789J1-001
10174	1636848	4	0564J1-001	10104	1638916	4	1793J1-001
10175	1636850	4	0566J1-001	10106	1638920	4	1792J1-001
10019	1637234	4	0791J1-001	10108	1638924	4	0001J1+000
10020	1637236	4	0793J1-001	10109	1638926	4	0003J1+000
10021	1637238	4	0795J1-001	10110	1638928	4	0005J1+000
10022	1637240	4	0797J1-001	10113	1638936	4	0013J1+000
10176	1637242	4	0799J1-001	10040	1638938	4	0015J1+000
10177	1637244	4	0801J1-001	10114	1638940	4	0017J1+000
10178	1637246	4	0803J1-001	10115	1638942	4	0019J1+000
10179	1637248	4	0805J1-001	10041	1638946	4	0023J1+000
10180	1637250	4	0807J1-001	10042	1638954	4	0031J1+000
10023	1637528	4	0965J1-001	10120	1638956	4	0033J1+000
10024	1637530	4	0967J1-001	10123	1639002	4	0039J1+000
10025	1637532	4	0969J1-001	10124	1639004	4	0041J1+000
10026	1637534	4	0971J1-001	10126	1639008	4	0045J1+000
10144	1637536	4	0973J1-001	10127	1639010	4	0047J1+000
10145	1637538	4	0975J1-001	10128	1639012	4	0049J1+000

IO

Table 8--continued

Rand Number	FDS Number	Camera[a]	Picture Number	Rand Number	FDS Number	Camera[a]	Picture Number
10129	1639014	4	0051J1+000	10074	1639148	4	0145J1+000
10130	1639016	4	0053J1+000	10075	1639150	4	0147J1+000
10053	1639024	4	0061J1+000	10076	1639152	4	0149J1+000
10132	1639026	4	0063J1+000	10077	1639154	4	0151J1+000
10133	1639028	4	0065J1+000	10078	1639156	4	0153J1+000
10134	1639030	4	0067J1+000	10079	1639158	4	0155J1+000
10054	1639036	4	0073J1+000	10080	1639200	4	0157J1+000
10055	1639038	4	0075J1+000	10155	1639216	3	0173J1+000
10056	1639040	4	0077J1+000	10156	1639218	3	0175J1+000
10057	1639042	4	0079J1+000	10157	1639220	3	0177J1+000
10136	1639044	4	0081J1+000	10158	1639222	3	0179J1+000
10043	1639046	4	0083J1+000	10159	1639224	3	0181J1+000
10044	1639048	4	0085J1+000	10160	1639226	3	0183J1+000
10143	1639049	3	0086J1+000	10161	1639235	3	0192J1+000
10045	1639050	4	0087J1+000	10162	1639237	3	0194J1+000
10046	1639052	4	0089J1+000	10163	1639239	3	0196J1+000
10047	1639054	4	0091J1+000	10170	1639241	3	0198J1+000
10058	1639056	4	0093J1+000	10164	1639243	3	0200J1+000
10151	1639057	3	0094J1+000	10171	1639257	3	0214J1+000
10048	1639058	4	0095J1+000	10165	1639259	3	0216J1+000
10050	1639102	4	0099J1+000	10166	1639301	3	0218J1+000
10051	1639104	4	0101J1+000	10167	1639315	3	0232J1+000
10052	1639106	4	0103J1+000	10168	1639317	3	0234J1+000
10059	1639108	4	0105J1+000	10537	2059213	2	1366J2-003
10061	1639112	4	0109J1+000	10503	2060805	2	0518J2-002
10062	1639114	4	0111J1+000	10504	2061530	2	0963J2-002
10063	1639116	4	0113J1+000	10501	2062133	2	1326J2-002
10064	1639118	4	0115J1+000	10502	2064152	2	0745J2-001
10065	1639120	4	0117J1+000	10530	2065942	2	0015J2+000
10066	1639122	4	0119J1+000	10531	2065944	2	0017J2+000
10067	1639124	4	0121J1+000	10532	2065946	2	0019J2+000
10068	1639126	4	0123J1+000	10533	2065948	2	0021J2+000
10069	1639128	4	0125J1+000	10534	2065950	2	0023J2+000
10070	1639130	4	0127J1+000	10507	2065952	2	0025J2+000
10152	1639133	3	0130J1+000	10535	2065954	2	0027J2+000
10169	1639135	3	0132J1+000	10553	2066422	2	0295J2+000
10153	1639137	3	0134J1+000	10554	2066438	2	0311J2+000
10072	1639144	4	0141J1+000	10555	2066527	2	0360J2+000
10154	1639145	3	0142J1+000	10506	2066912	2	0585J2+000
10073	1639146	4	0143J1+000				

[a]Camera Code: 1 = Voyager 2 Wide Angle
2 = Voyager 2 Narrow Angle
3 = Voyager 1 Wide Angle
4 = Voyager 1 Narrow Angle

4 = Voyager 1 narrow angle. Table 9 gives the same data for Europa, Table 10 for Ganymede, and Table 11 for Callisto.

Figures 4 through 12 identify some of the control points on maps and pictures of Io. Table 12 gives the coordinates of the control points on Io.

Eight volcanic plumes were observed on Io during the encounters. The plumes are most easily seen above the limb or terminator in the pictures but it is difficult to identify the sources of the eruption from vertical views. Estimates of the locations of the plume sources have been made and their coordinates computed; they are listed in Table 13.

Figures 13 through 21 identify some of the control points on maps and pictures of Europa. Table 14 gives the coordinates of the control points on Europa.

Figures 22 through 28 identify some of the control points on maps and pictures of Ganymede. Table 15 gives the coordinates of the control points on Ganymede.

Figures 29 through 36 identify some of the control points on maps and pictures of Callisto. Table 16 gives the coordinates of the control points on Callisto.

Table 9

EUROPA: PICTURES IN THE CONTROL NET

Rand Number	FDS Number	Camera[a]	Picture Number	Rand Number	FDS Number	Camera[a]	Picture Number
20021	1631255	4	0810J1-003	20531	2064931	2	1204J2-001
20022	1632318	4	1433J1-003	20520	2064934	2	1207J2-001
20025	1633456	4	0331J1-002	20508	2064937	2	1210J2-001
20027	1634220	4	0775J1-002	20528	2064940	2	1213J2-001
20029	1634918	4	1193J1-002	20529	2064943	2	1216J2-001
20031	1635208	4	1364J1-002	20521	2064946	2	1219J2-001
20033	1635707	4	1663J1-002	20509	2064949	2	1222J2-001
20034	1635711	4	1667J1-002	20532	2064952	2	1225J2-001
20035	1636027	4	0063J1-001	20533	2064955	2	1228J2-001
20037	1636900	4	0576J1-001	20522	2064958	2	1231J2-001
20501	2058607	2	1000J2-003	20510	2065001	2	1234J2-001
20502	2059317	2	1430J2-003	20534	2065004	2	1237J2-001
20503	2060513	2	0346J2-002	20536	2065007	2	1240J2-001
20504	2061215	2	0768J2-002	20511	2065139	2	1332J2-001
20537	2062516	2	1549J2-002	20512	2065143	2	1336J2-001
20505	2062524	2	1557J2-002	20513	2065151	2	1344J2-001
20518	2064910	2	1183J2-001	20514	2065155	2	1348J2-001
20506	2064913	2	1186J2-001	20515	2065159	2	1352J2-001
20526	2064916	2	1189J2-001	20516	2065203	2	1356J2-001
20527	2064919	2	1192J2-001	20517	2065207	2	1360J2-001
20519	2064922	2	1195J2-001	20523	2065211	2	1364J2-001
20507	2064925	2	1198J2-001	20524	2065215	2	1368J2-001
20530	2064928	2	1201J2-001	20525	2065219	2	1372J2-001

[a]Camera Code: 1 = Voyager 2 Wide Angle
2 = Voyager 2 Narrow Angle
3 = Voyager 1 Wide Angle
4 = Voyager 1 Narrow Angle

Table 10

GANYMEDE: PICTURES IN THE CONTROL NET

Rand Number	FDS Number	Camera[a]	Picture Number	Rand Number	FDS Number	Camera[a]	Picture Number
30023	1626439	4	1514J1-005	30049	1640315	3	0832J1+000
30026	1628936	4	1211J1-004	30067	1640316	4	0833J1+000
30029	1629944	4	0019J1-003	30069	1640318	4	0835J1+000
30030	1630824	4	0539J1-003	30050	1640322	4	0839J1+000
30033	1634229	4	0784J1-002	30073	1640324	4	0841J1+000
30034	1634233	4	0788J1-002	30074	1640326	4	0843J1+000
30035	1635655	4	1651J1-002	30075	1640328	4	0845J1+000
30036	1635659	4	1655J1-002	30051	1640332	4	0849J1+000
30038	1640142	4	0739J1+000	30077	1640334	4	0851J1+000
30039	1640144	4	0741J1+000	30078	1640336	4	0853J1+000
30119	1640148	4	0745J1+000	30079	1640338	4	0855J1+000
30040	1640152	4	0749J1+000	30052	1640342	4	0859J1+000
30122	1640154	4	0751J1+000	30057	1640344	4	0861J1+000
30124	1640158	4	0755J1+000	30081	1640346	4	0863J1+000
30041	1640202	4	0759J1+000	30082	1640348	4	0865J1+000
30126	1640204	4	0761J1+000	30053	1640352	4	0869J1+000
30128	1640208	4	0765J1+000	30058	1640354	4	0871J1+000
30042	1640212	4	0769J1+000	30088	1640358	4	0875J1+000
30131	1640214	4	0771J1+000	30061	1640719	4	1076J1+000
30134	1640218	4	0775J1+000	30062	1640723	4	1080J1+000
30044	1640222	4	0779J1+000	30063	1640725	4	1082J1+000
30138	1640224	4	0781J1+000	30501	2056843	2	1756J2-004
30140	1640228	4	0785J1+000	30502	2058258	2	0811J2-003
30045	1640232	4	0789J1+000	30503	2060811	2	0524J2-002
30142	1640234	4	0791J1+000	30530	2063059	2	0092J2-001
30144	1640238	4	0795J1+000	30504	2063101	2	0094J2-001
30047	1640252	4	0809J1+000	30531	2063103	2	0096J2-001
30152	1640254	4	0811J1+000	30532	2063105	2	0098J2-001
30154	1640256	4	0813J1+000	30533	2063107	2	0100J2-001
30156	1640258	4	0815J1+000	30505	2063109	2	0102J2-001
30159	1640302	4	0819J1+000	30562	2063111	2	0104J2-001
30160	1640304	4	0821J1+000	30534	2063113	2	0106J2-001
30161	1640306	4	0823J1+000	30535	2063115	2	0108J2-001
30162	1640308	4	0825J1+000	30536	2063117	2	0110J2-001
30048	1640312	4	0829J1+000	30506	2063119	2	0112J2-001
30066	1640314	4	0831J1+000				

[a]Camera Code: 1 = Voyager 2 Wide Angle
2 = Voyager 2 Narrow Angle
3 = Voyager 1 Wide Angle
4 = Voyager 1 Narrow Angle

Table 11

CALLISTO: PICTURES IN THE CONTROL NET

Rand Number	FDS Number	Camera[a]	Picture Number	Rand Number	FDS Number	Camera[a]	Picture Number
40001	1596105	4	1300J1-015	40147	1642143	4	0140J1+001
40004	1598442	4	0917J1-014	40149	1642147	4	0144J1+001
40007	1600653	4	0448J1-013	40151	1642151	4	0148J1+001
40009	1600657	4	0452J1-013	40153	1642155	4	0152J1+001
40011	1602404	4	1479J1-013	40155	1642159	4	0156J1+001
40014	1604414	4	0889J1-012	40157	1642203	4	0160J1+001
40016	1604418	4	0893J1-012	40159	1642207	4	0164J1+001
40019	1606530	4	0365J1-011	40161	1642211	4	0168J1+001
40020	1608056	4	1291J1-011	40163	1642215	4	0172J1+001
40022	1608100	4	1295J1-011	40165	1642219	4	0176J1+001
40026	1612251	4	0206J1-009	40166	1642222	4	0179J1+001
40028	1612255	4	0210J1-009	40167	1642225	4	0182J1+001
40035	1615533	4	0368J1-008	40098	1642426	4	0303J1+001
40037	1617415	4	1490J1-008	40099	1642428	4	0305J1+001
40039	1619419	4	0894J1-007	40100	1642430	4	0307J1+001
40041	1621257	4	0212J1-006	40101	1642432	4	0309J1+001
40043	1625311	4	0826J1-005	40102	1642434	4	0311J1+001
40047	1630803	4	0518J1-003	40103	1642436	4	0313J1+001
40048	1630807	4	0522J1-003	40104	1642438	4	0315J1+001
40049	1632159	4	1354J1-003	40105	1642440	4	0317J1+001
40051	1632304	4	1419J1-003	40106	1642442	4	0319J1+001
40054	1639944	4	0621J1+000	40107	1642444	4	0321J1+001
40055	1639950	4	0627J1+000	40108	1642446	4	0323J1+001
40058	1639956	4	0633J1+000	40109	1642448	4	0325J1+001
40060	1641800	4	1717J1+000	40110	1642450	4	0327J1+001
40061	1641806	4	1723J1+000	40111	1642452	4	0329J1+001
40062	1641814	4	1731J1+000	40112	1642456	4	0333J1+001
40063	1641822	4	1739J1+000	40113	1642458	4	0335J1+001
40065	1641830	4	1747J1+000	40090	1642500	4	0337J1+000
40066	1641838	4	1755J1+000	40091	1642502	4	0339J1+001
40067	1641846	4	1763J1+000	40092	1642504	4	0341J1+001
40068	1641854	4	1771J1+000	40093	1642506	4	0343J1+001
40069	1641902	4	1779J1+000	40080	1642508	4	0345J1+001
40138	1642118	4	0115J1+001	40081	1642510	4	0347J1+001
40139	1642121	4	0118J1+001	40082	1642514	4	0351J1+001
40141	1642127	4	0124J1+001	40083	1642516	4	0353J1+001
40142	1642130	4	0127J1+001	40084	1642518	4	0355J1+001
40143	1642133	4	0130J1+001	40085	1642520	4	0357J1+001
40144	1642136	4	0133J1+001	40086	1642522	4	0359J1+001
40145	1642139	4	0136J1+001	40087	1642524	4	0361J1+001

CALLISTO

Table 11--Continued

Rand Number	FDS Number	Camera[a]	Picture Number
40147	1642143	4	0140J1+001
40149	1642147	4	0144J1+001
40151	1642151	4	0148J1+001
40153	1642155	4	0152J1+001
40155	1642159	4	0156J1+001
40157	1642203	4	0160J1+001
40159	1642207	4	0164J1+001
40076	1642805	4	0522J1+001
40077	1642807	4	0524J1+001
40078	1642809	4	0526J1+001
40547	2045330	2	0243J2-007
40548	2047922	2	1795J2-007
40549	2049504	2	0937J2-006
40550	2050705	2	1658J2-006
40501	2052127	2	0720J2-005

[a]Camera Code: 1 = Voyager 2 Wide Angle
2 = Voyager 2 Narrow Angle
3 = Voyager 1 Wide Angle
4 = Voyager 1 Narrow Angle

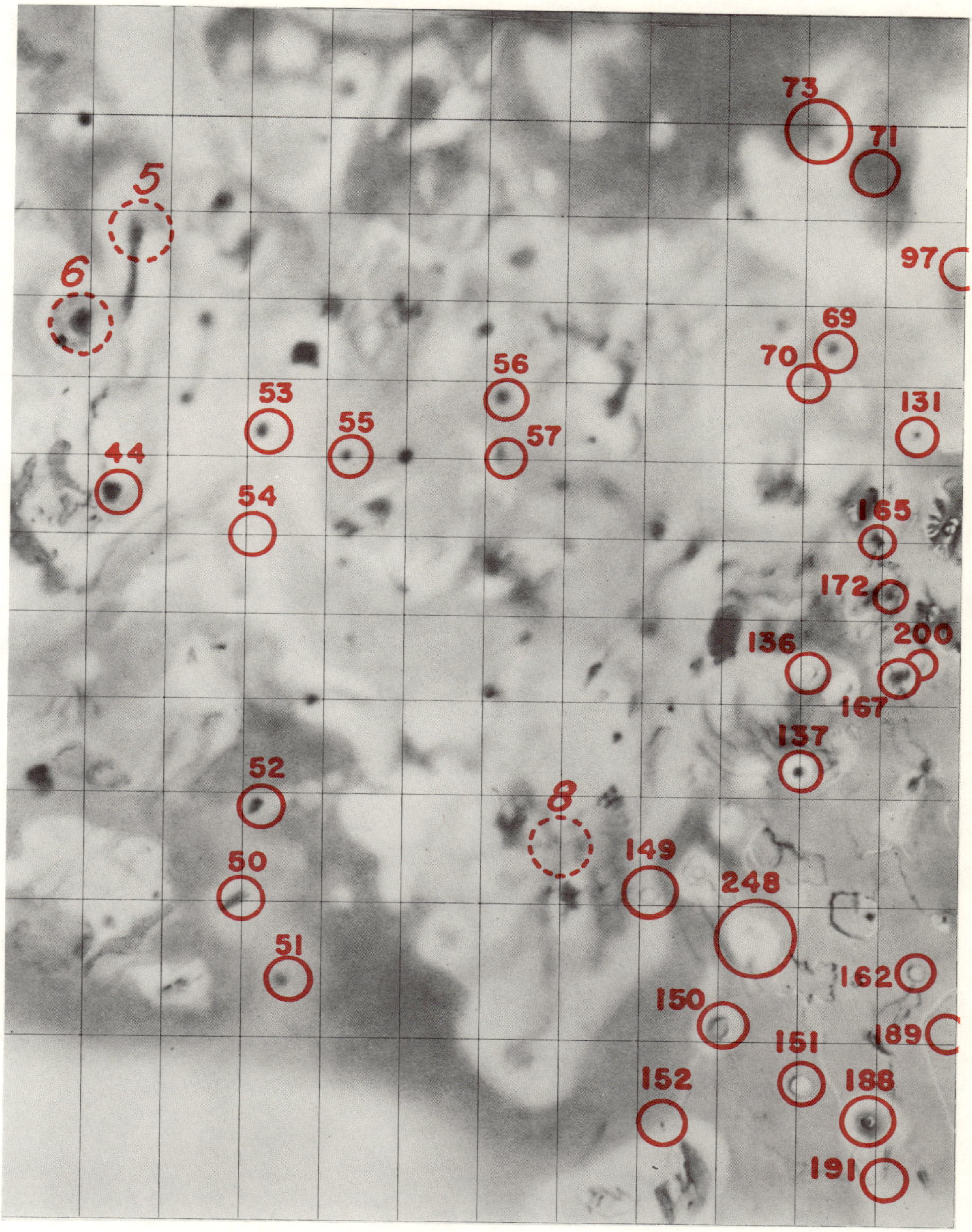

Fig. 4 — Io: Mercator map with control points identified in the region of the prime meridian and east

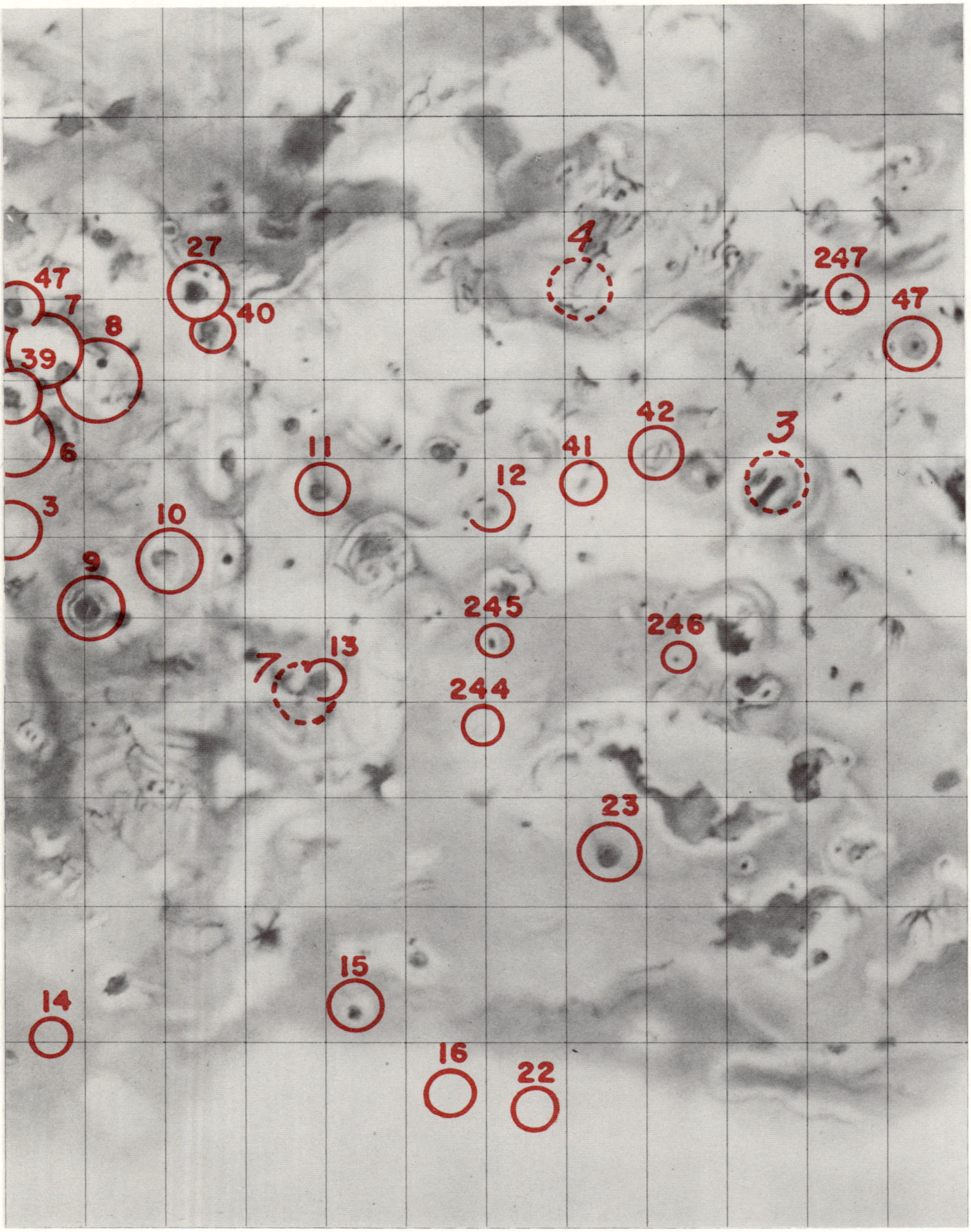

Fig. 5 — Io: Mercator map with control points identified in the region of 180° longitude

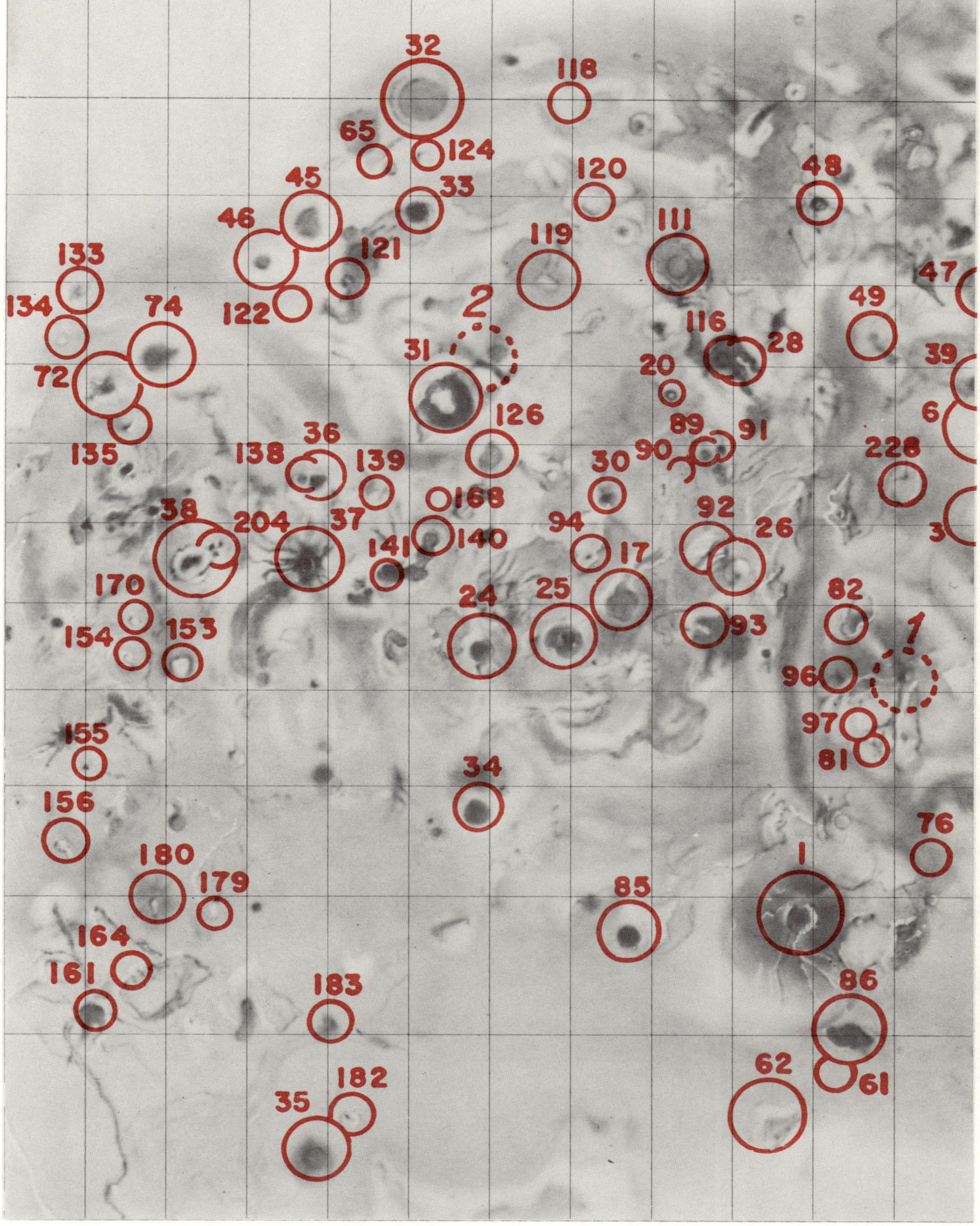

Fig. 6 — Io: Mercator map with control points identified in the region of the prime meridian and west

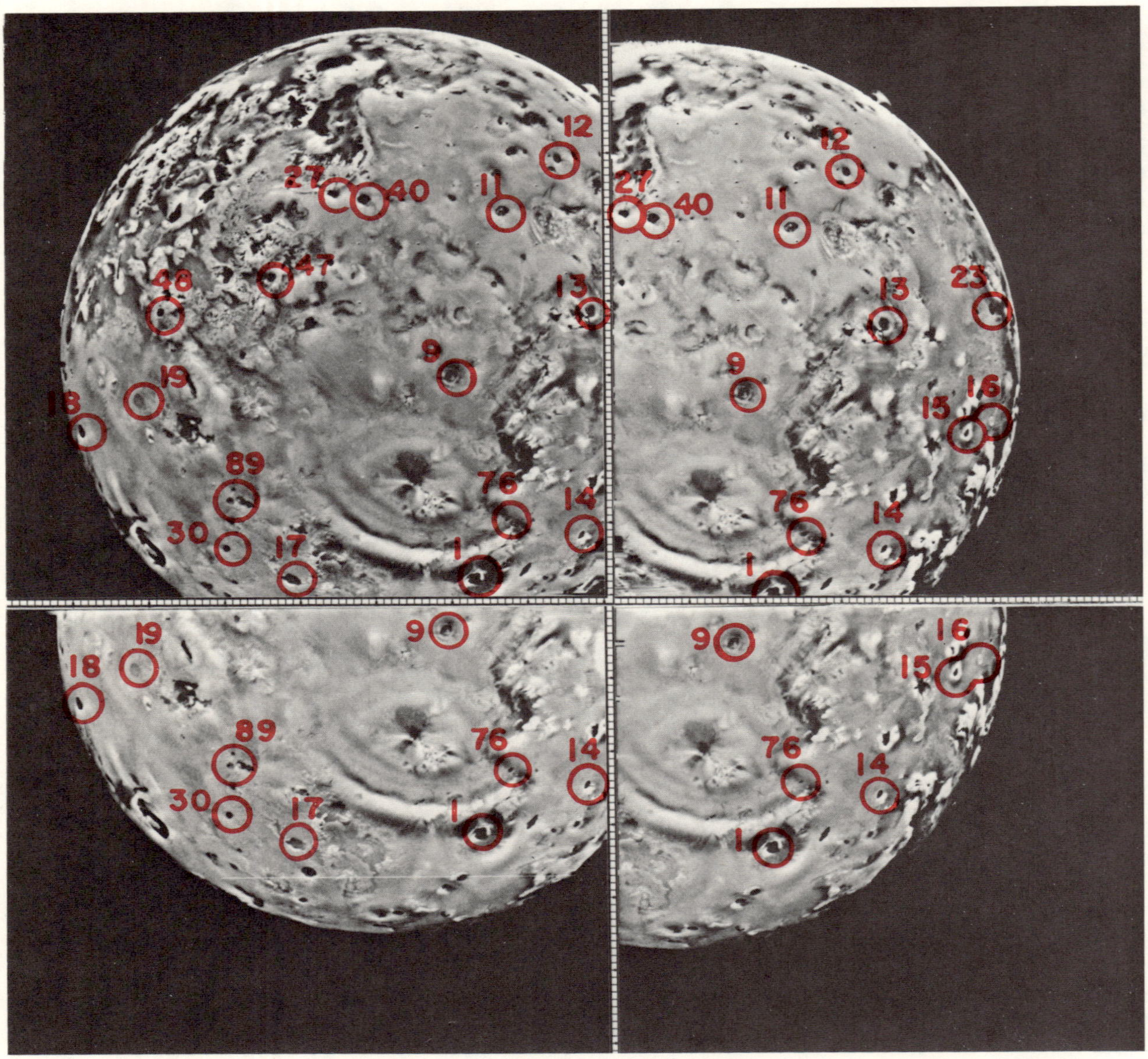

Fig. 7 — Io: Four picture mosaic with control points identified
(FDS 16377.50, 16377.52, 16377.54, 16377.56)

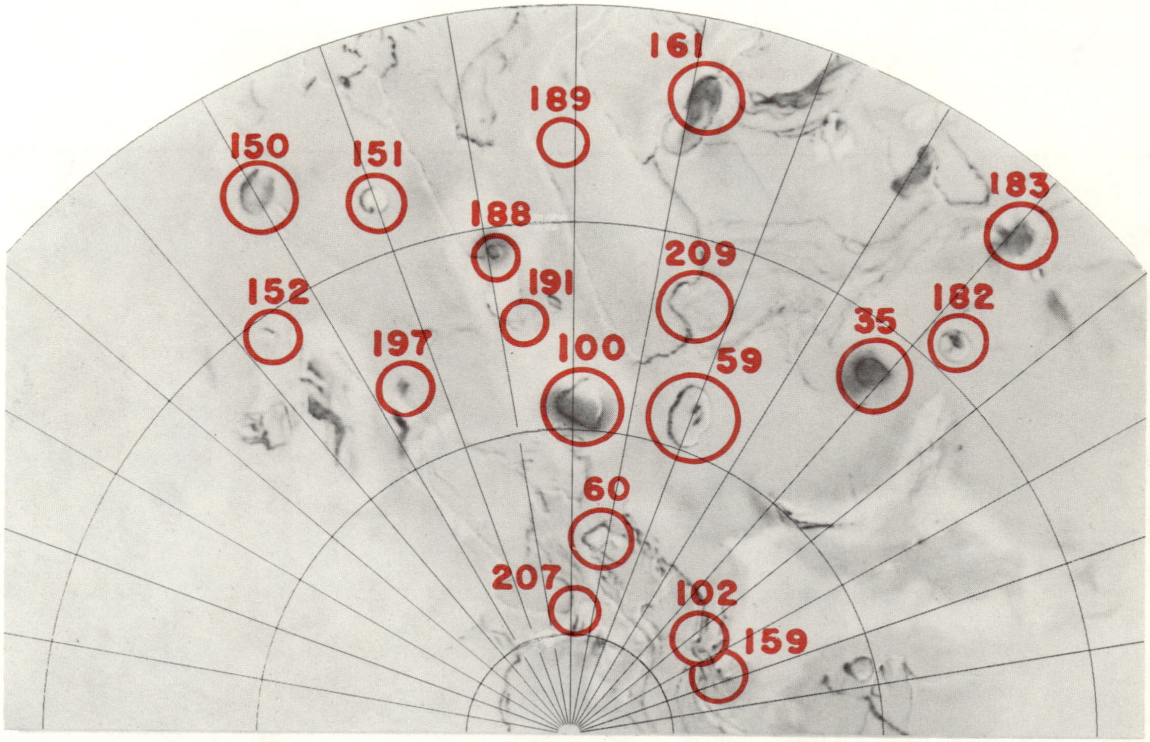

Fig. 8 — Io: Stereographic map with control points identified in the region of the south pole

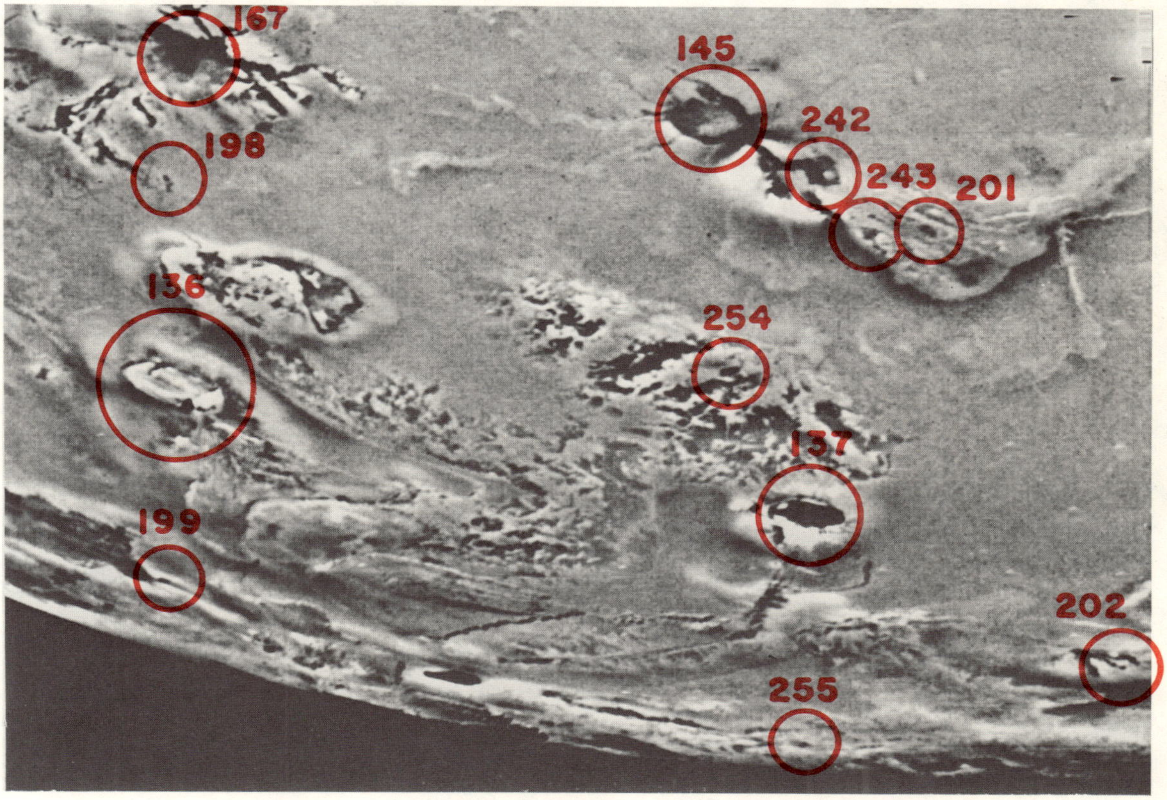

Fig. 9 — Io: Near-encounter picture with control points identified

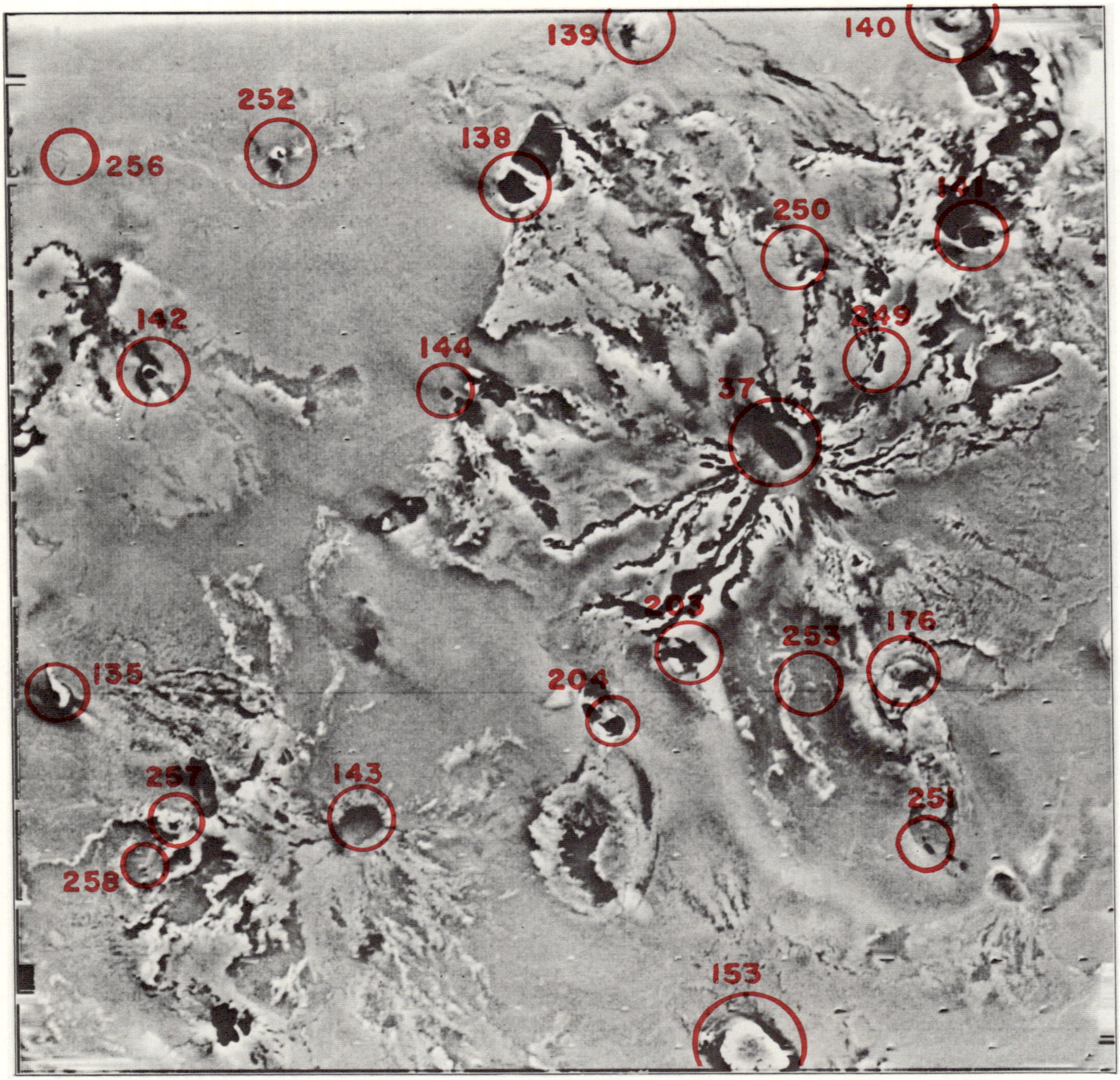

Fig. 10 — Io: Near-encounter picture with control points identified

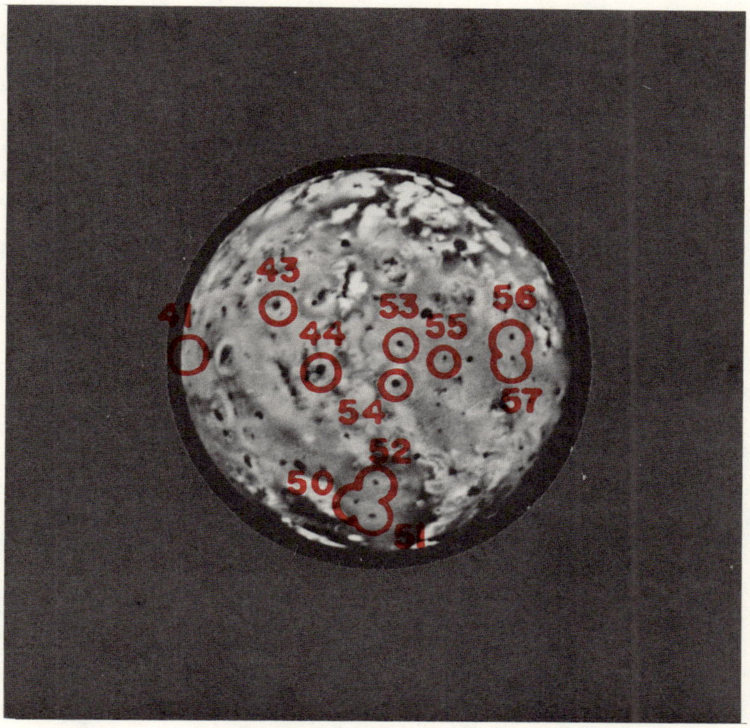

Fig. 11 — Io: Far-encounter picture with control points identified

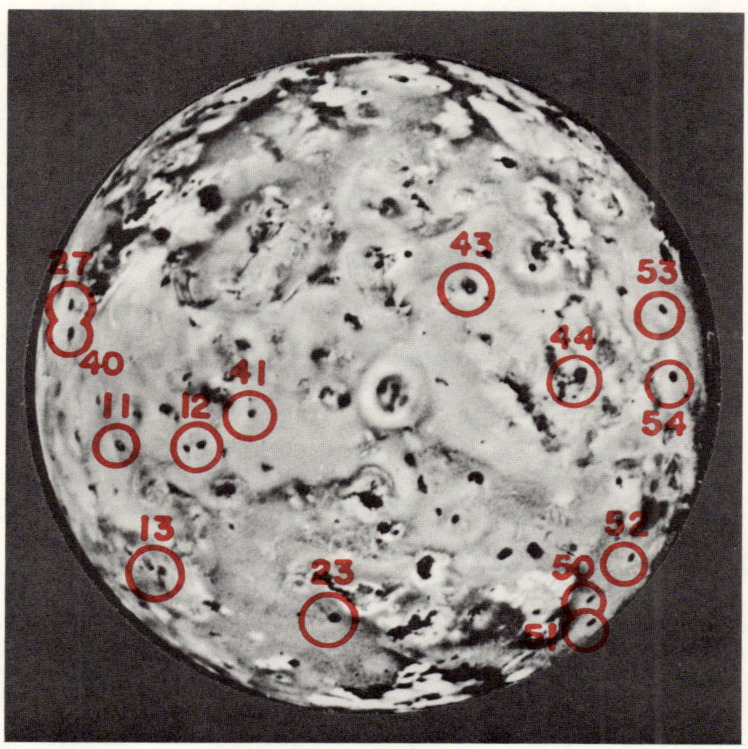

Fig. 12 — Io: Far-encounter picture with control points identified

Table 12

IO: COORDINATES OF CONTROL POINTS

(degrees)

Point	Lat.	Long.	Point	Lat.	Long.	Point	Lat.	Long.
1	-39.6	272.2	41	-1.7	176.6	85	-40.6	287.8
2	-16.9	254.5	42	1.3	167.2	86	-48.3	266.9
3	-4.6	244.1	43	14.7	136.0	88	-58.8	267.1
4	17.7	265.2	44	-3.4	118.9	89	6.1	280.1
5	17.5	271.5	45	35.3	322.1	90	3.5	283.8
6	7.6	242.0	46	28.3	328.0	91	6.4	277.2
7	15.7	240.2	47	22.3	239.7	92	-5.6	279.1
8	12.0	232.8	48	36.8	260.8	93	-12.7	279.0
9	-13.5	235.9	49	18.5	254.7	94	-4.9	292.8
10	-8.9	225.9	50	-47.1	107.1	95	-20.2	277.4
11	-3.7	205.2	51	-53.4	98.8	96	-20.1	261.9
12	-5.6	187.0	52	-38.7	99.8	97	-23.2	260.7
13	-27.1	207.6	53	6.0	96.5	98	-10.6	270.9
14	-55.6	251.6	54	-5.6	97.0	99	-43.5	239.7
15	-55.9	204.6	55	3.6	83.7	100	-68.1	349.4
16	-59.6	189.2	56	11.5	61.9	101	-72.3	321.8
17	-10.8	288.5	57	5.4	61.9	102	-70.5	296.3
18	37.7	307.5	58	-34.0	134.9	103	-66.3	342.7
19	30.3	279.1	59	-65.0	330.5	104	-79.2	13.0
20	15.4	282.1	60	-73.5	338.8	105	-79.7	14.2
21	39.6	287.3	61	-54.1	269.6	111	30.3	279.1
22	-59.0	177.7	62	-62.0	281.5	112	30.2	264.6
23	-42.4	173.5	63	-76.5	282.1	113	29.7	264.2
24	-16.3	305.9	64	57.1	320.9	114	15.4	281.5
25	-14.8	295.1	65	48.7	310.3	115	15.9	281.5
26	-7.2	275.8	66	45.2	337.7	116	19.0	274.7
27	22.1	219.9	67	44.4	349.2	117	6.4	277.1
28	19.1	271.9	68	43.4	357.7	118	53.5	285.8
29	16.6	303.8	69	19.2	24.0	119	28.6	292.6
30	1.9	290.3	70	15.9	27.5	120	39.6	287.4
31	13.2	309.7	71	25.2	10.4	121	26.9	316.5
32	54.4	301.7	72	22.5	351.4	122	21.6	323.5
33	37.6	306.9	73	35.9	15.4	123	52.0	357.7
34	-32.6	304.4	74	14.2	341.4	124	47.3	302.7
35	-56.9	312.4	75	24.7	336.2	125	15.7	313.3
36	1.9	322.2	76	-38.0	254.6	126	6.0	303.7
37	-8.4	325.3	80	-31.4	268.3	127	11.1	322.1
38	-9.7	339.9	81	-24.4	260.5	128	-3.0	304.1
39	11.8	242.7	82	-15.2	261.2	129	26.2	34.6
40	16.9	218.0	83	-16.3	272.1	130	51.9	2.8

IO

Table 12--Continued

Point	Lat.	Long.	Point	Lat.	Long.	Point	Lat.	Long.
131	9.1	7.1	173	-6.6	34.2	219	-75.1	359.6
132	3.6	18.6	174	-4.1	41.2	220	-70.9	44.3
133	16.3	352.8	176	-14.7	328.9	221	-70.4	47.8
134	10.1	349.3	177	-19.5	324.4	224	-55.0	349.9
135	5.7	344.7	178	-20.1	325.3	225	-62.8	332.4
136	-22.0	26.0	179	-40.4	334.7	226	-64.3	333.0
137	-35.4	26.7	180	-40.0	341.0	227	-71.7	327.7
138	1.8	323.7	182	-53.1	307.6	228	0.5	255.2
139	1.7	317.8	183	-48.1	311.3	229	0.2	267.6
140	-5.5	312.2	184	-73.6	265.5	230	-4.2	269.0
141	-9.5	317.2	185	-75.1	248.1	233	-67.0	247.9
142	7.9	334.6	186	-69.4	279.5	234	-62.7	245.1
143	-3.9	342.1	187	-62.0	352.0	235	-66.9	249.3
144	0.3	329.7	188	-62.9	6.7	236	-1.8	264.1
145	-31.7	7.9	189	-56.7	357.2	237	-18.2	305.1
146	-6.7	251.2	190	-61.7	12.6	243	-35.6	11.1
147	-47.2	13.1	193	-83.1	47.5	244	-31.6	187.3
148	-47.4	40.7	194	-71.8	53.3	245	-20.9	186.2
149	-49.7	46.3	195	-55.7	346.9	246	-22.9	166.1
150	-58.6	36.5	196	-35.3	238.0	247	22.2	145.2
151	-61.2	22.2	197	-70.2	28.9	248	-51.9	31.0
152	-64.5	49.9	198	-21.4	16.9	259	-71.7	35.0
153	-16.5	342.1	199	-17.4	54.2	260	-70.8	36.2
154	-16.0	348.9	200	-20.5	7.9	261	-70.4	36.4
155	-28.9	354.4	201	-36.7	10.6	262	-68.4	20.1
156	-38.3	355.5	202	-41.4	38.9	263	-64.2	31.8
157	-74.5	314.0	203	-9.4	331.9	264	-69.1	44.5
158	-79.1	320.0	204	-8.7	335.1	265	-65.7	46.5
159	-69.8	292.6	205	-70.5	302.4	266	-64.8	48.2
160	-2.3	297.8	206	-72.5	313.6	267	-64.2	49.7
161	-51.4	343.9	207	-79.0	342.0	268	-63.0	47.9
162	-49.9	3.0	208	-76.4	329.1	269	-73.8	36.0
163	-45.5	0.7	209	-60.9	337.6	270	-67.2	43.2
164	-46.9	340.9	211	-63.2	348.8	271	-66.1	23.9
165	-2.6	15.9	212	-64.4	355.5	272	30.1	221.6
166	-18.8	3.6	213	-65.0	359.6	273	4.5	209.0
167	-21.1	12.0	215	-82.0	30.5	274	43.1	249.0
168	-0.3	309.9	216	-80.5	35.7	275	15.3	231.6
169	-11.7	305.6	217	-78.3	2.8	276	11.8	206.3
170	-11.2	348.8	218	-73.5	354.3	277	35.3	137.1

IO

Table 12--Continued

Point	Lat.	Long.	Point	Lat.	Long.	Point	Lat.	Long.
278	37.0	118.0	302	19.8	62.3	336	-13.0	282.1
280	16.4	123.3	303	-25.3	63.3	337	-14.7	284.4
281	-25.8	144.7	305	40.1	46.3	338	-13.9	287.9
282	-41.5	137.0	306	41.6	56.8	339	-13.0	285.8
283	-48.2	123.9	307	38.2	87.4	340	-11.6	281.7
284	-46.1	117.8	308	40.9	74.5	341	-9.3	283.2
285	-52.6	193.2	309	-47.7	69.9	342	-8.3	274.9
286	43.3	190.8	310	4.6	75.8	343	-10.7	274.6
287	64.9	190.0	311	-12.7	75.2	5131	8.9	4.6
288	60.8	142.6	312	-27.6	87.4	5132	3.0	16.3
289	58.7	118.2	314	-4.1	39.8	5137	-35.6	25.6
290	40.7	192.1	315	-43.8	54.0	5145	-31.5	9.2
291	7.8	145.3	316	10.8	110.1	5148	-47.6	39.3
292	-12.6	138.1	317	19.7	108.4	5151	-61.3	20.9
293	17.2	192.6	318	-17.2	42.2	5152	-64.9	47.1
294	67.5	248.1	319	-30.3	73.4	5153	-16.7	343.7
295	19.8	176.4	320	22.2	89.5	5154	-16.0	350.6
296	54.6	260.2	321	-29.6	45.5	5155	-28.7	355.6
297	49.5	265.0	322	-45.2	52.7	5165	-3.0	13.8
298	54.4	24.8	323	-5.8	83.9	5170	-11.2	350.5
299	53.1	13.2	324	-41.7	78.4	5203	-9.3	333.8
300	-47.9	36.8	335	-12.4	276.5	5204	-8.5	336.9
301	27.4	57.5						

Table 13

IO: COORDINATES OF THE ERUPTIVE CENTERS
(degrees)

Plume Number	Name	Latitude	Longitude
1	Pele	-19.4	256.8
2	Loki	19.0	305.3
3	Prometheus	-2.9	153.0
4	Volund	21.5	177.0
5	Amirani	27.2	118.7
6	Maui	18.9	122.4
7	Marduk	-27.9	209.7
8	Masubi	-45.2	52.7

NOTE: Eruptive centers are identified on Figs. 4, 5, and 6 by dashed circles.

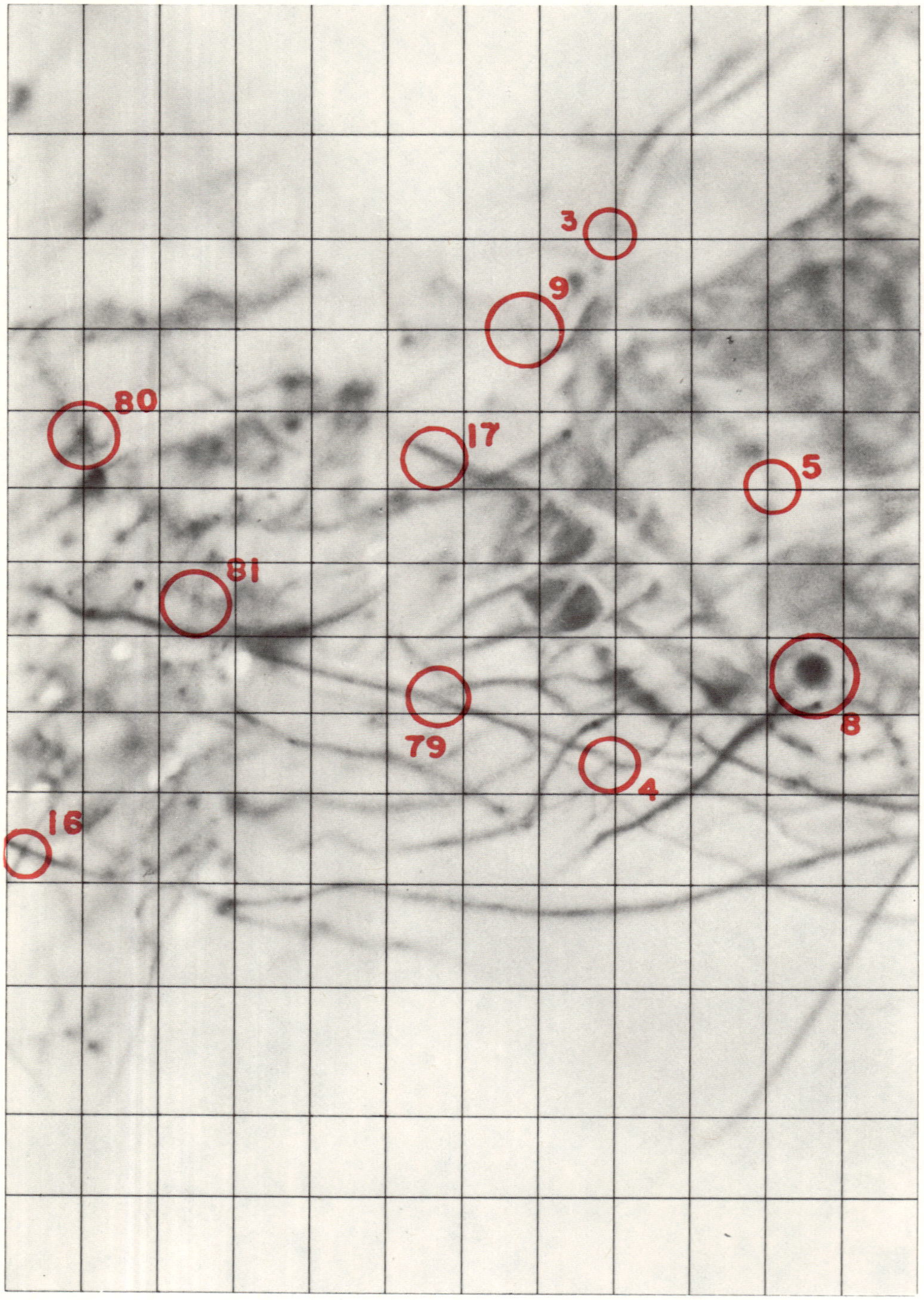

Fig. 13 — Europa: Mercator map with control points identified in the region of the prime meridian

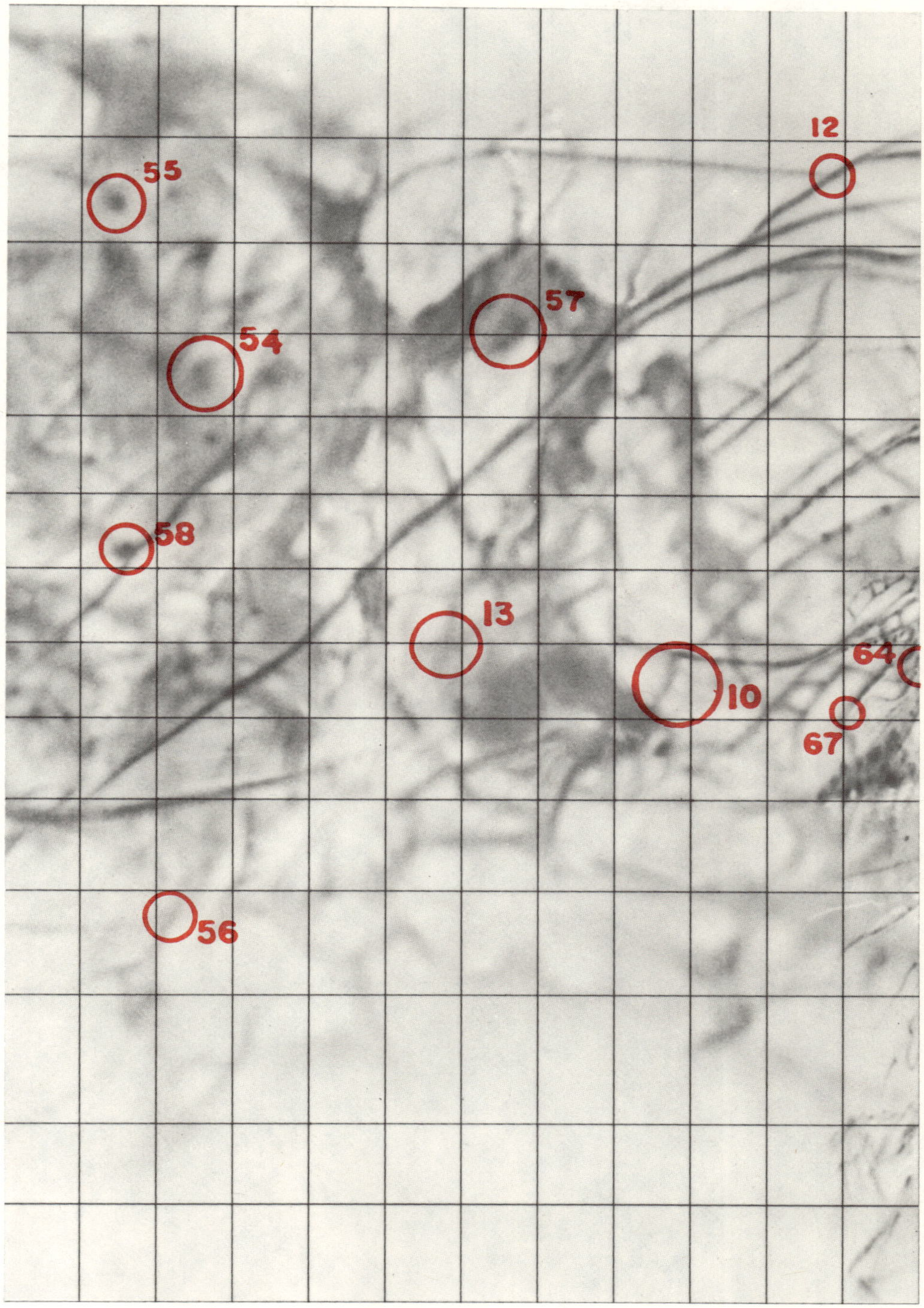

Fig 14 — Europa: Mercator map with control points identified in the region east of the prime meridian

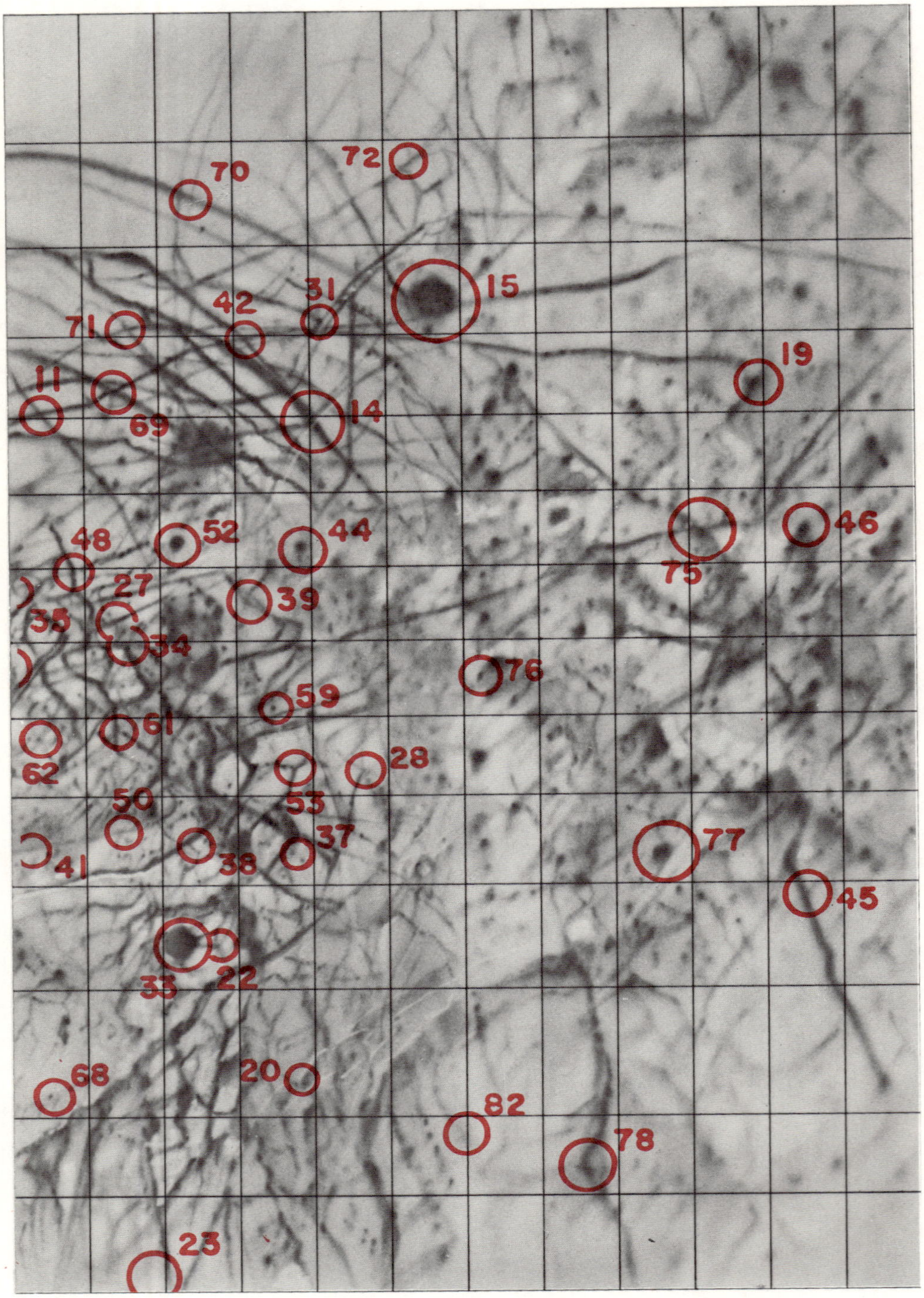

Fig. 15 — Europa: Mercator map with control points identified in the region west of the prime meridian

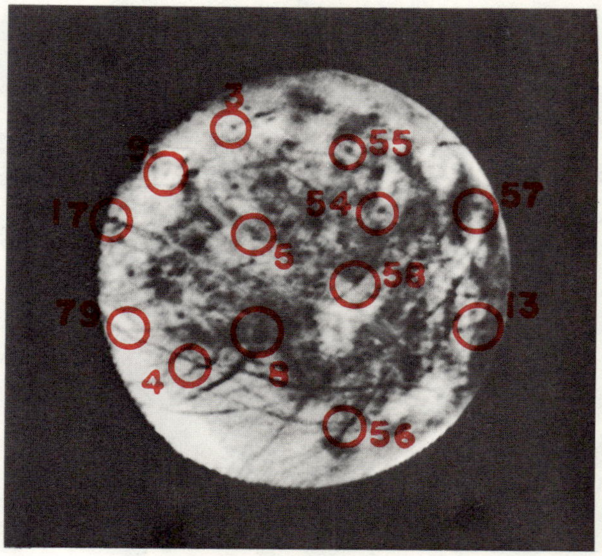

Fig. 16 — Europa: Far-encounter picture with control points identified

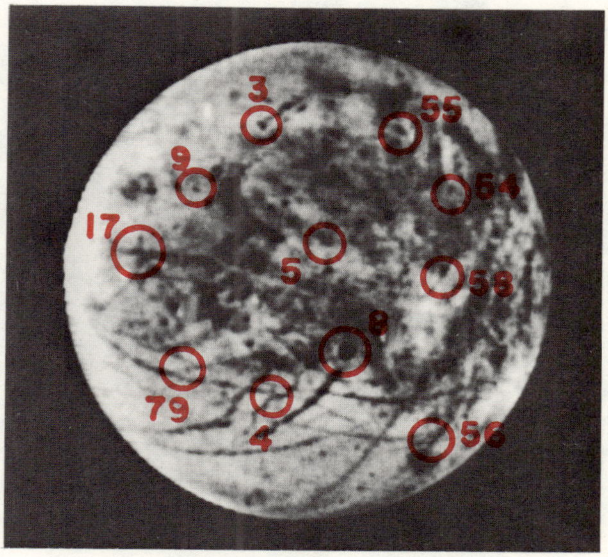

Fig. 17 — Europa: Far-encounter picture, overlapping with Fig. 16, with control points identified

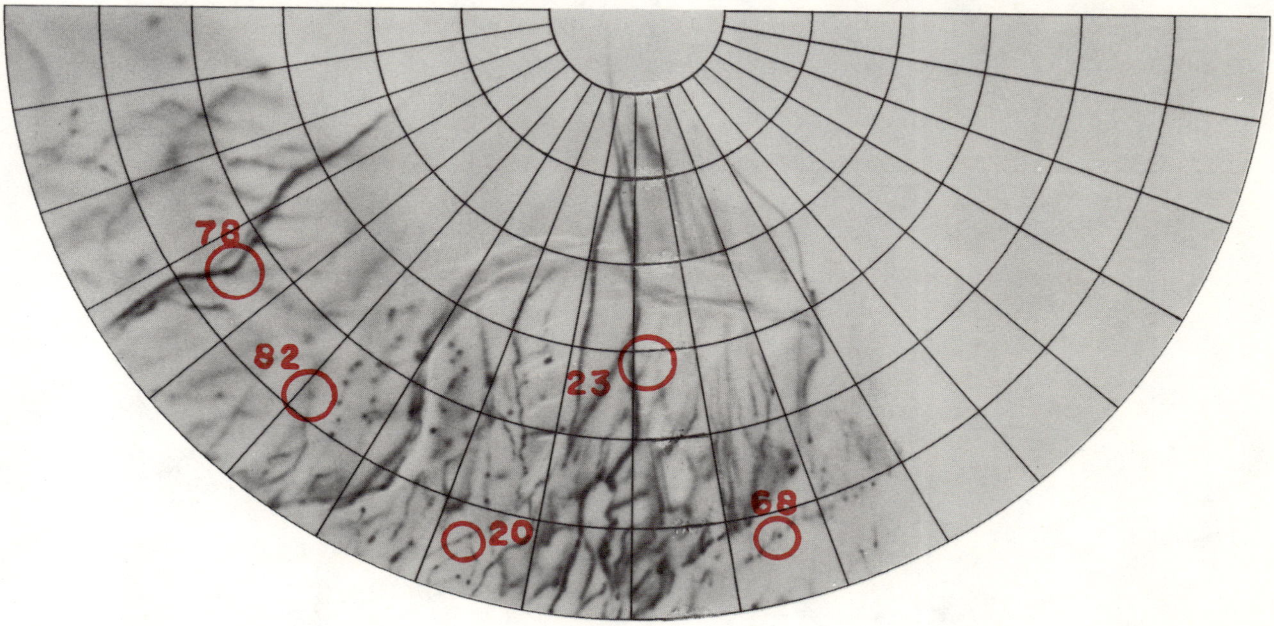

Fig. 18 — Europa: Stereographic map with control points identified in the region of the south pole

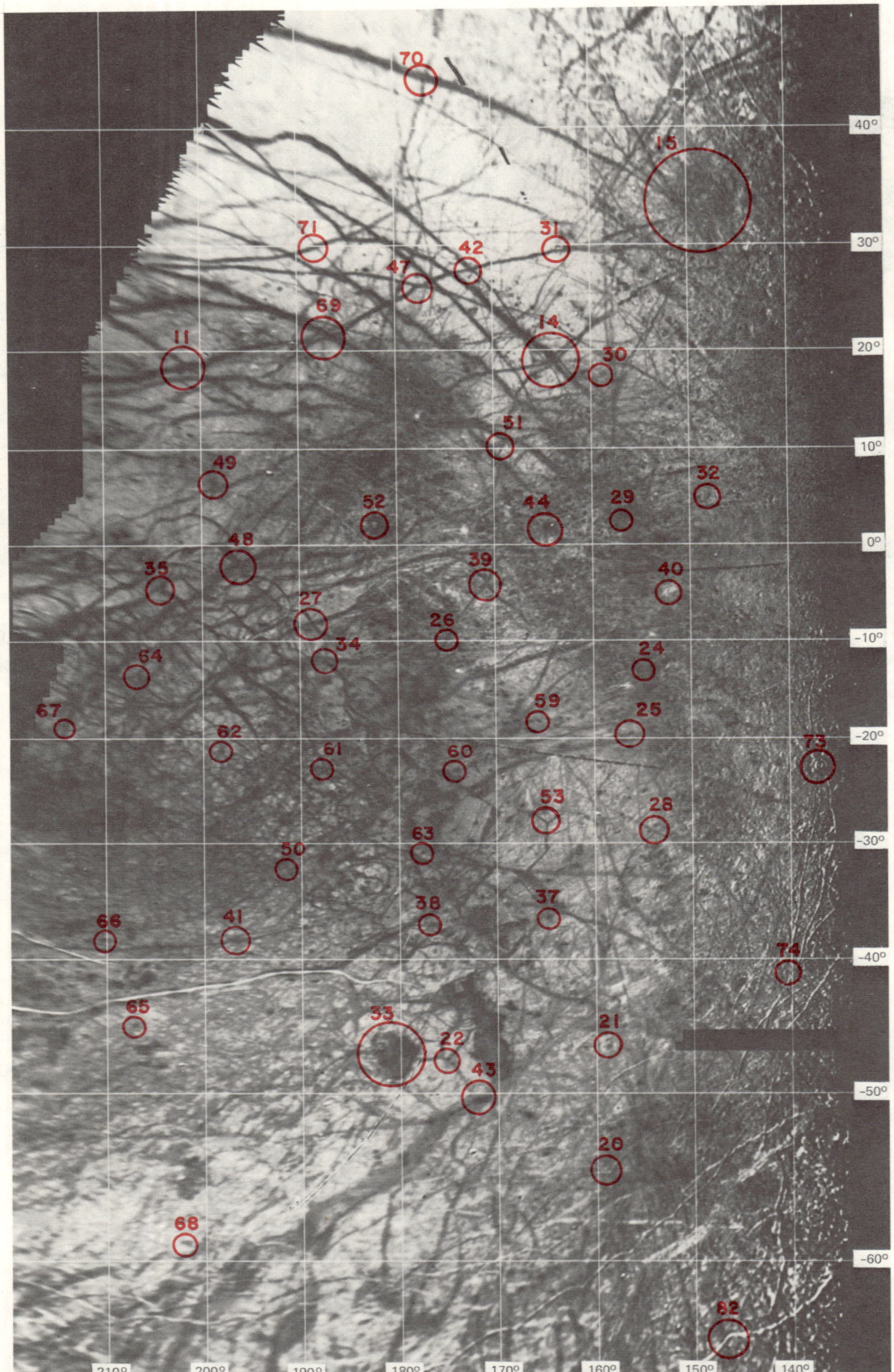

Fig. 19 — Europa: Mercator projection, computer mosaic with control points identified (computer mosaic by Joel A. Mosher, Image Processing Laboratory, JPL)

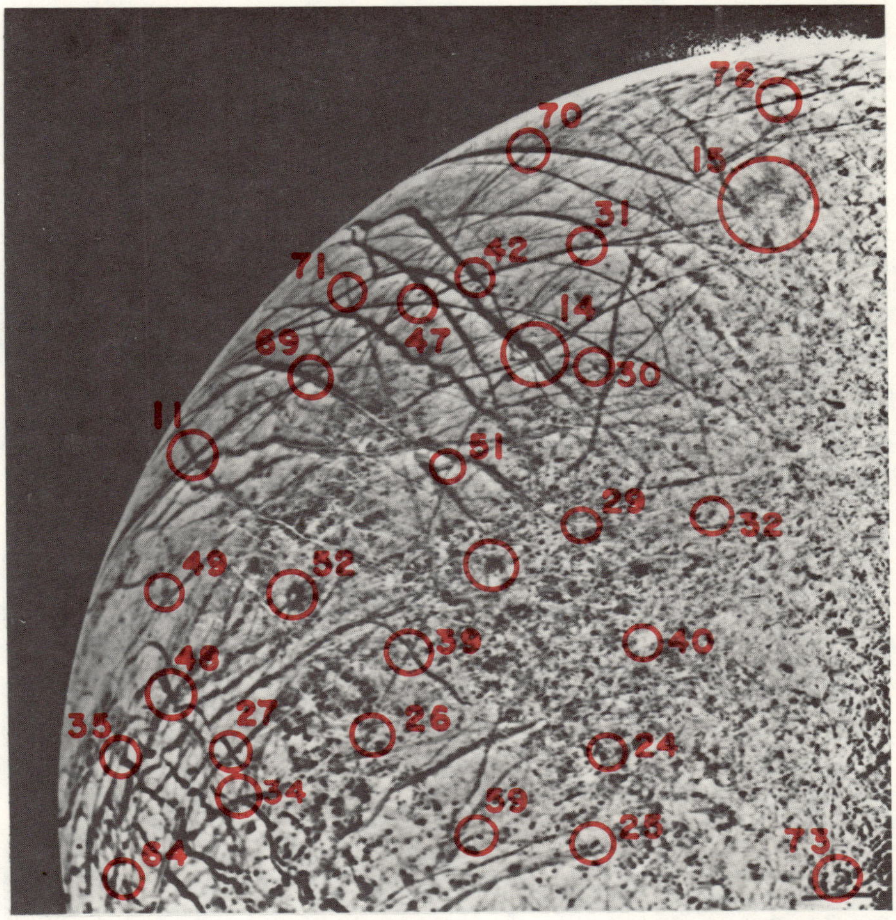

Fig. 20 — Europa: Near-encounter picture with control points identified

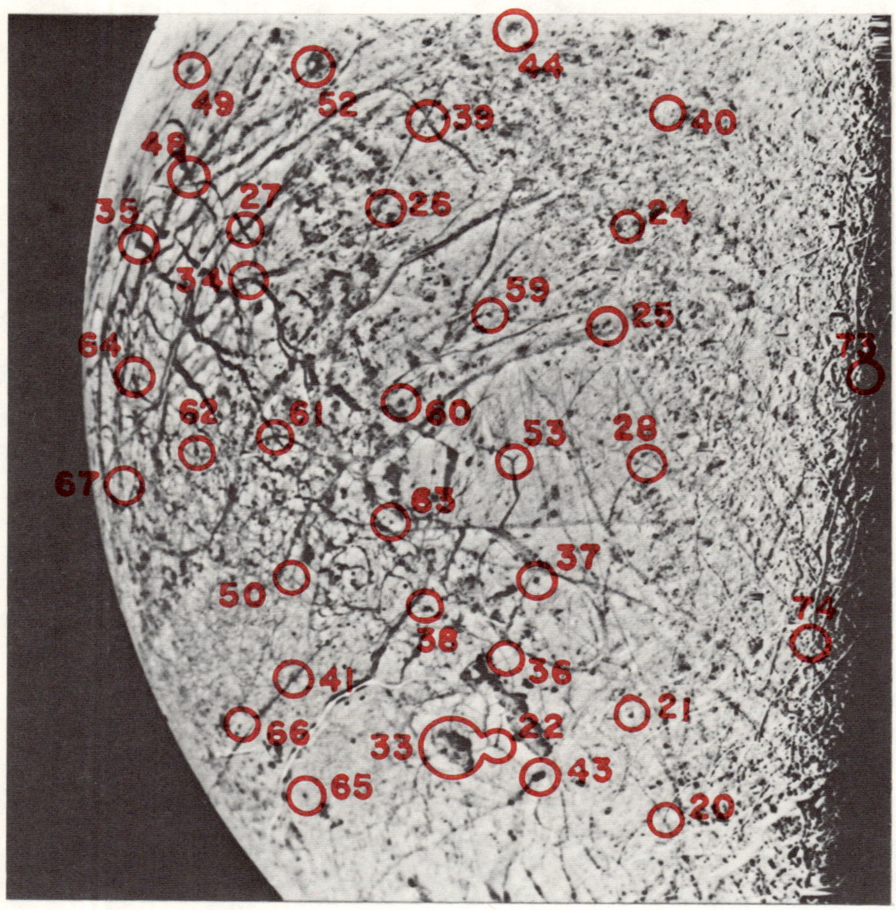

Fig. 21 — Europa: Near-encounter picture with control points identified

Table 14

EUROPA: COORDINATES OF CONTROL POINTS

Point	Lat.	Long.	Point	Lat.	Long.	Point	Lat.	Long.
3	43.7	357.3	35	-4.6	205.5	69	20.8	186.7
4	-25.0	352.9	36	-42.1	171.3	70	42.8	176.2
5	12.1	336.7	37	-36.8	165.7	71	29.5	188.2
8	-14.9	333.0	38	-37.2	178.5	72	47.1	148.8
9	25.3	10.5	39	-4.4	171.2	73	-23.5	137.5
10	-8.1	236.0	40	-5.8	152.4	74	-42.3	141.1
11	18.2	202.0	41	-38.4	198.6	82	-64.5	149.0
12	48.0	221.3	42	27.1	172.2	2001	-28.5	314.5
13	-3.2	265.8	43	-50.8	173.6	2002	30.5	253.5
14	18.4	163.8	44	1.2	164.4	2003	-25.3	339.9
15	32.5	147.2	45	-30.6	92.3	2004	14.9	227.7
16	-26.3	75.2	46	10.8	95.2	2006	48.0	220.3
17	11.1	25.0	47	25.5	177.3	2007	31.2	149.3
18	46.0	18.0	48	-2.4	196.9	2008	-18.8	205.1
19	30.8	103.2	49	6.4	199.5	2009	-1.2	171.4
20	-55.7	160.6	50	-32.3	191.9	2010	-25.2	195.9
21	-47.3	160.1	51	9.7	169.3	2011	10.5	110.0
22	-48.0	176.8	52	1.8	182.6	2013	-34.8	175.8
23	-69.0	211.4	53	-28.6	165.6	2015	-40.7	153.6
24	-13.5	155.2	59	-18.6	166.2	2016	-35.5	120.9
25	-20.2	156.7	60	-23.4	174.8	2017	-56.2	111.7
26	-10.2	174.7	61	-23.0	188.6	2018	-20.1	94.0
27	-8.5	189.3	62	-21.7	199.7	2019	-46.8	82.1
28	-29.6	154.4	63	-31.0	178.5	2020	-57.6	177.7
29	4.1	157.1	64	-13.2	208.1	2021	-75.5	222.0
30	16.9	158.6	65	-44.8	209.8	2022	-61.6	183.9
32	4.0	147.9	66	-38.0	213.1	2023	-49.7	147.0
33	-47.5	182.1	67	-18.3	216.5	2025	-58.3	154.4
34	-12.1	188.7	68	-58.8	205.7			

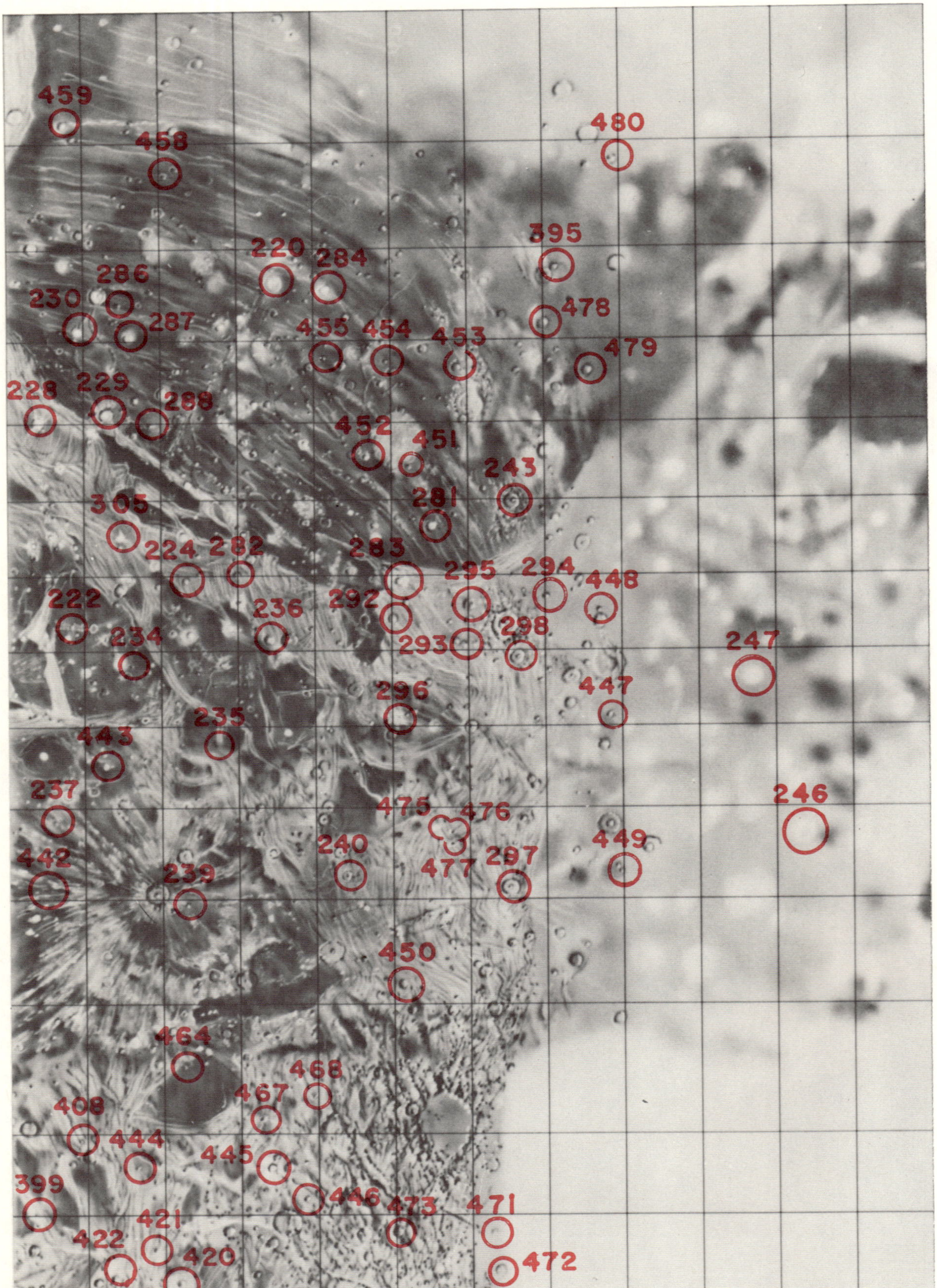

Fig. 22 — Ganymede: Mercator map with control points identified in the region west of the prime meridian

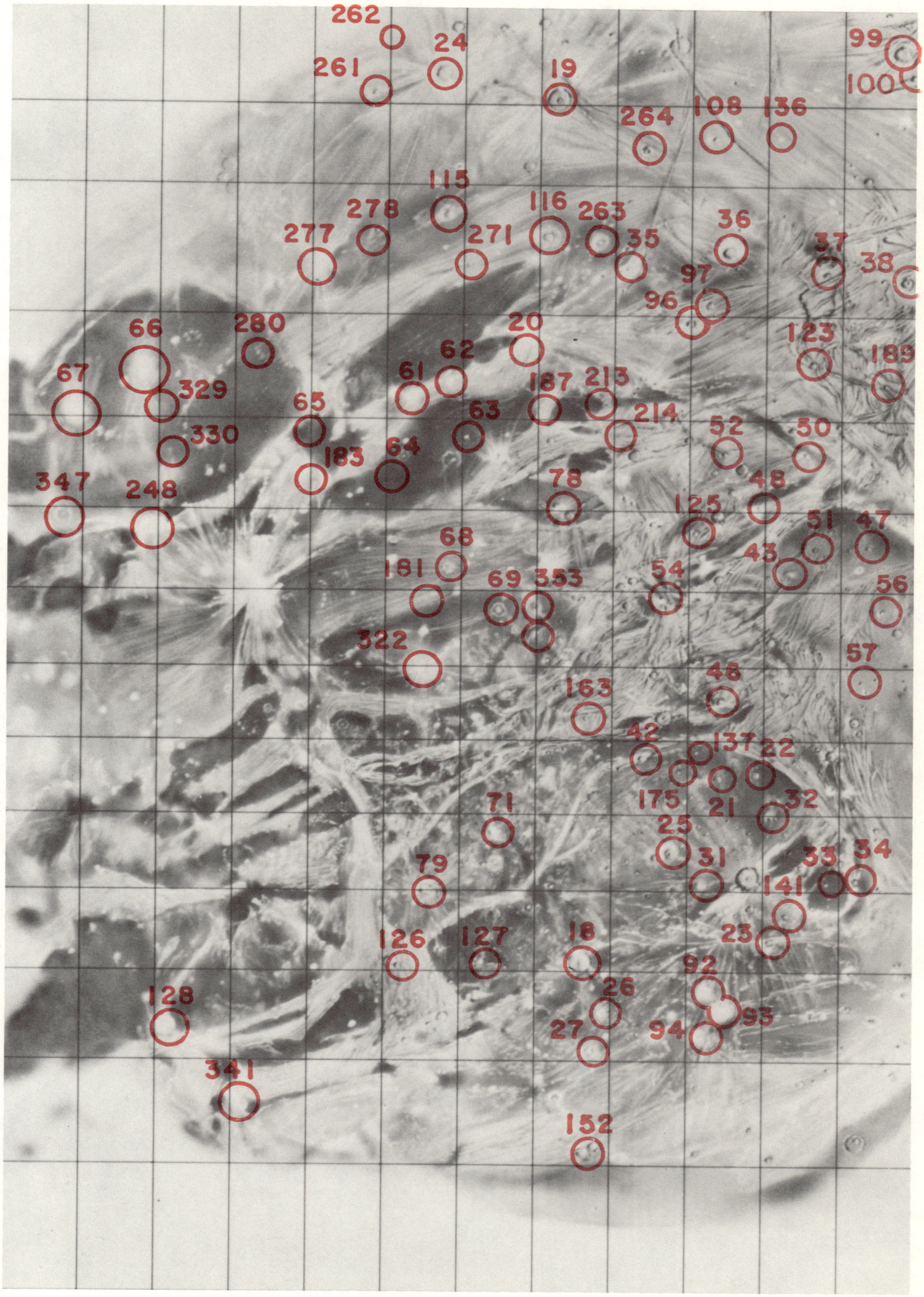

Fig. 23 — Ganymede: Mercator map with control points identified in the region of the prime meridian

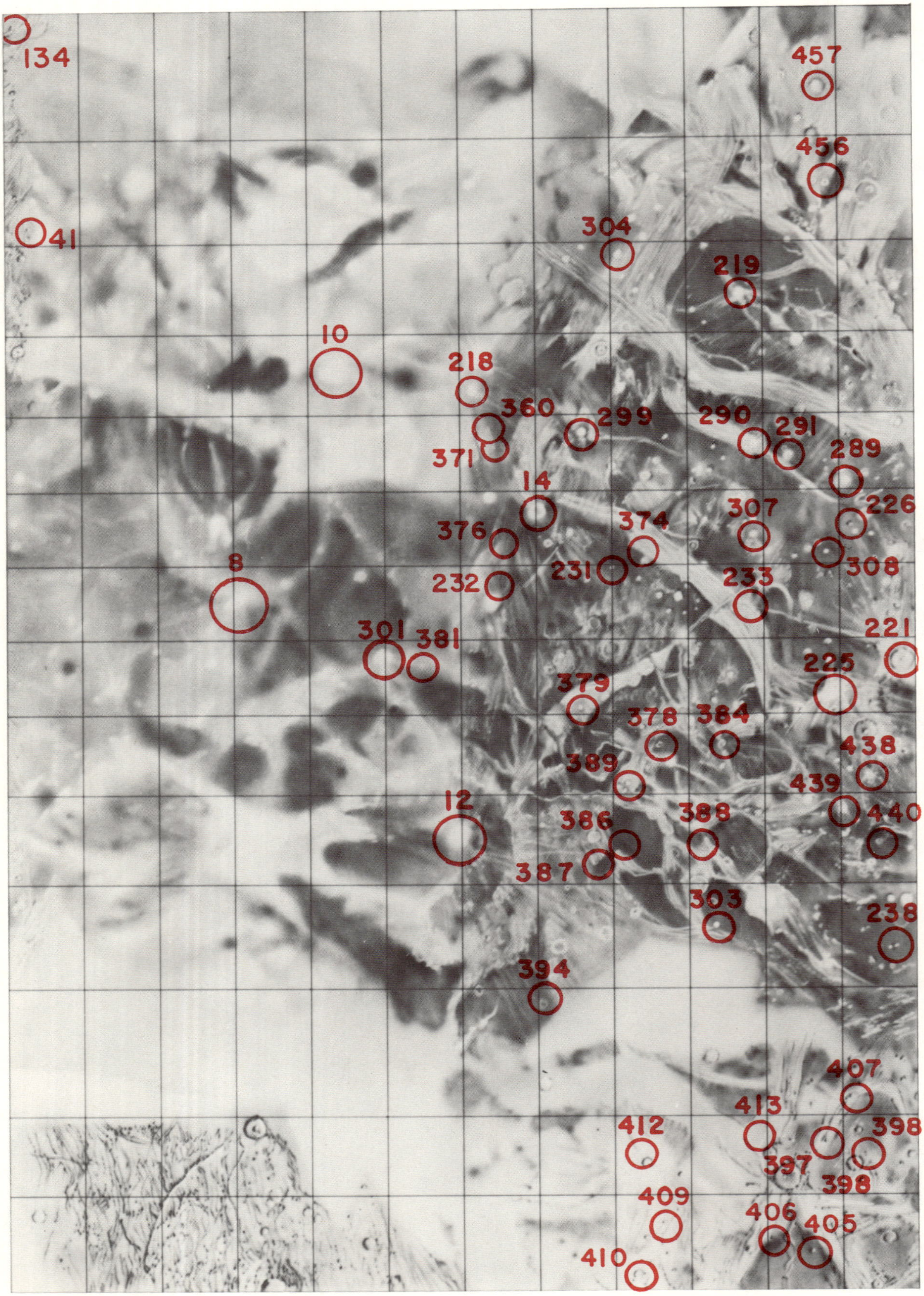

Fig. 24 — Ganymede: Mercator map with control points identified in the region east of the prime meridian

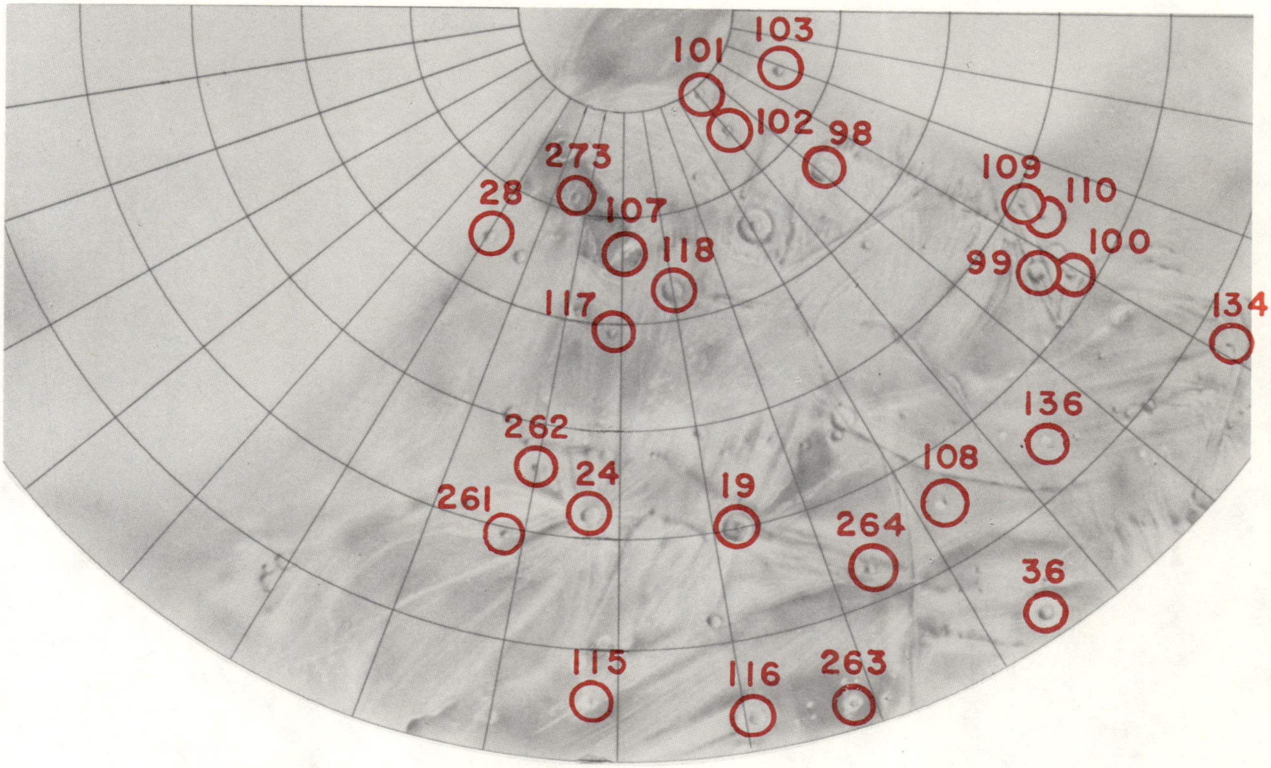

Fig. 25 — Ganymede: Stereographic map with control points identified in the region of the north pole

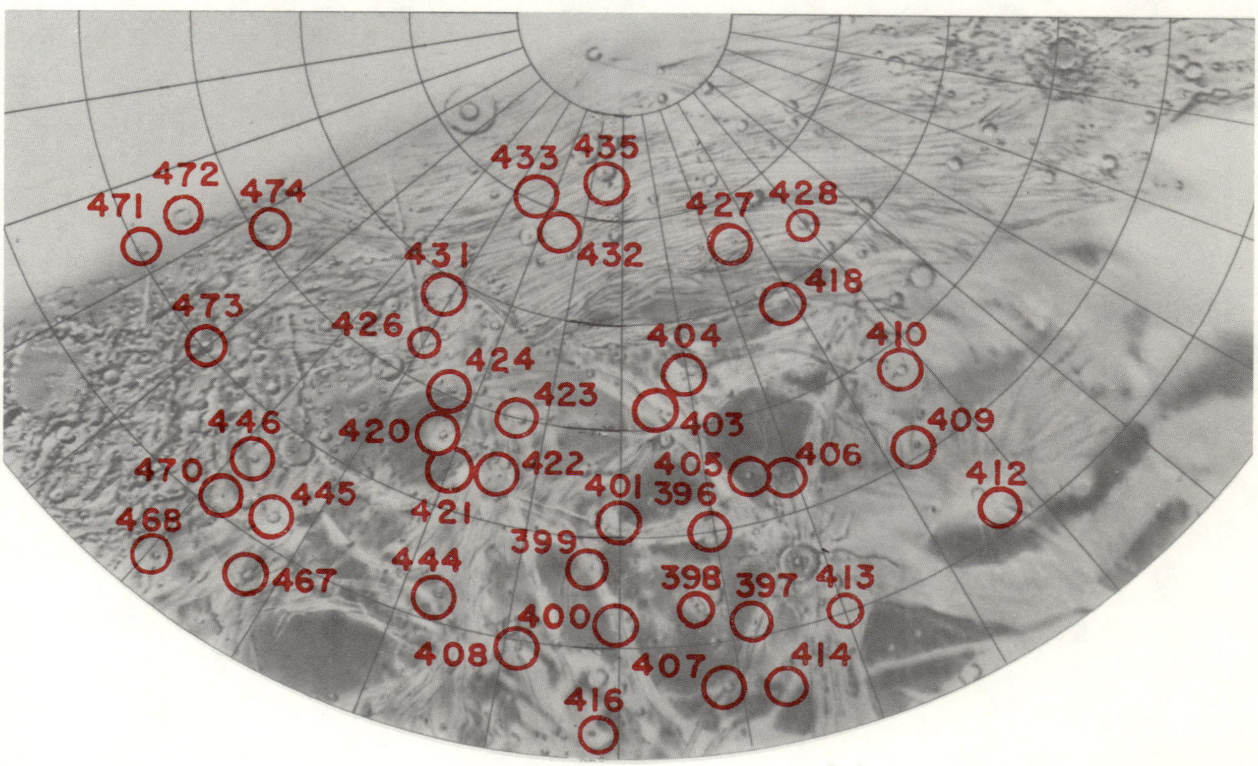

Fig. 26 — Ganymede: Stereographic map with control points identified in the region of the south pole

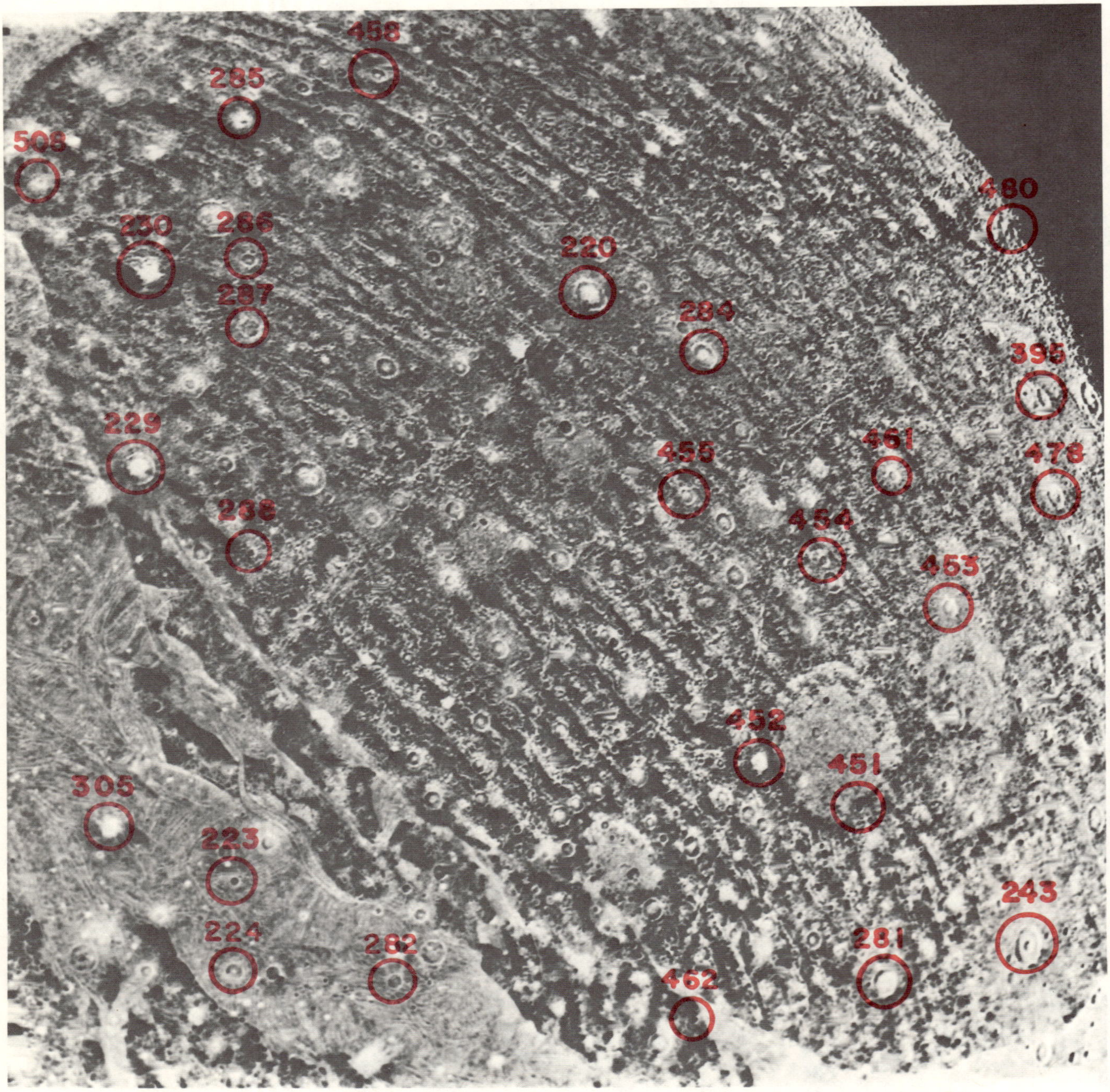

Fig. 27 — Ganymede: Near-encounter picture with control points identified

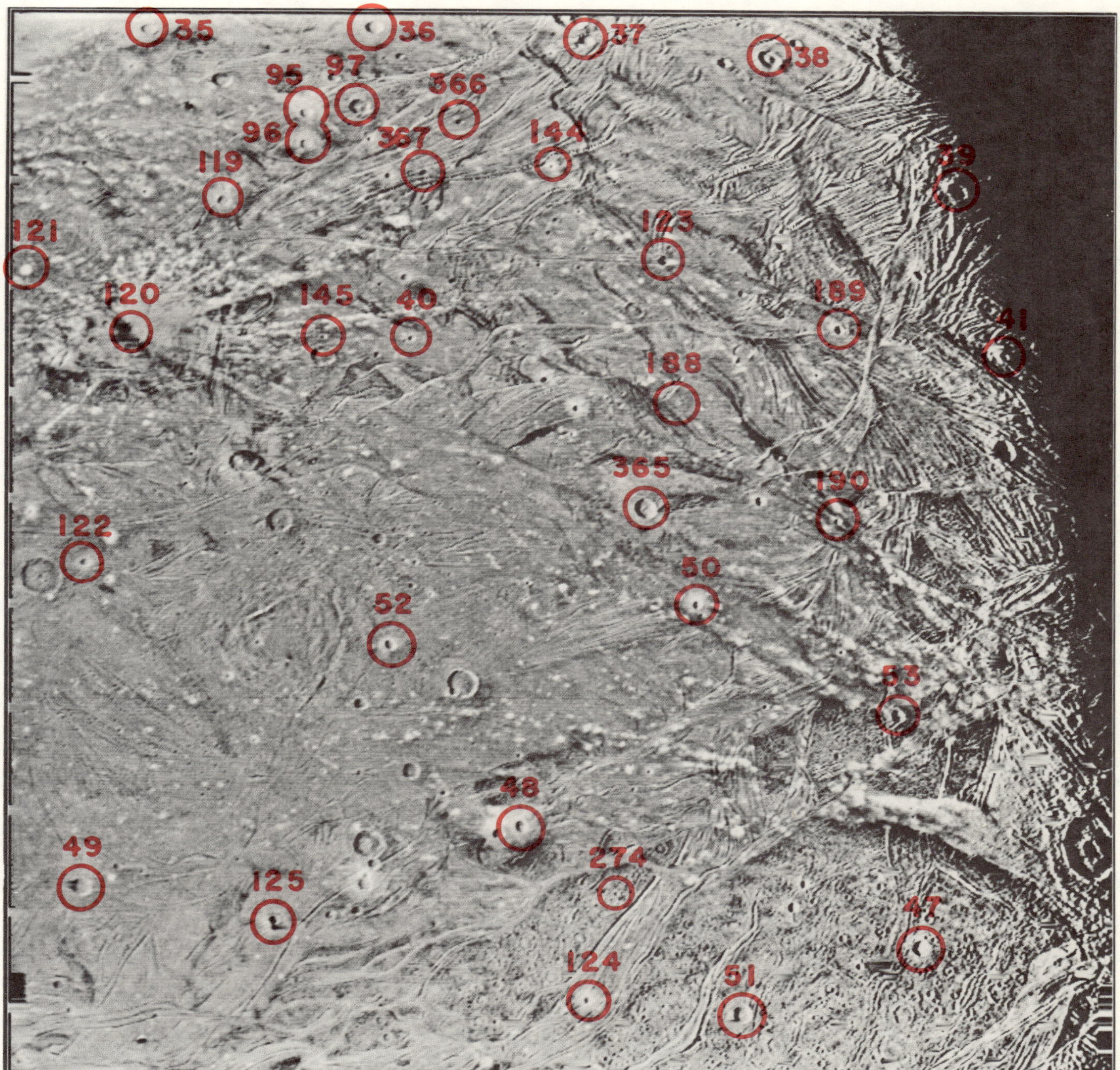

Fig. 28 — Ganymede: Near-encounter picture with control points identified

Table 15

GANYMEDE: COORDINATES OF CONTROL POINTS

Point	Lat.	Long.	Point	Lat.	Long.	Point	Lat.	Long.
8	-7.0	280.2	50	34.9	324.5	99	63.4	331.8
9	14.4	273.4	51	23.8	323.2	100	62.4	329.5
10	33.4	262.2	52	33.4	333.7	101	72.8	7.5
12	-23.2	242.5	53	31.8	318.6	102	71.7	4.7
13	7.3	293.5	54	11.2	338.8	103	76.5	348.0
14	17.4	226.7	61	38.2	16.6	104	78.6	23.4
15	49.5	343.2	62	38.9	10.3	105	79.8	10.7
16	0.7	14.8	63	31.1	10.1	106	80.7	16.7
17	15.6	333.7	64	28.0	18.8	109	67.8	328.4
18	-37.6	350.2	65	34.0	29.9	110	66.6	326.8
19	61.4	3.6	66	41.5	54.8	112	-17.6	45.5
20	42.8	0.5	68	14.4	9.2	113	-25.7	71.8
21	-13.3	332.6	69	7.7	2.3	115	55.2	14.6
22	-12.0	328.2	70	1.6	20.5	116	52.8	2.5
23	-35.3	329.8	71	-23.0	358.7	117	68.1	24.1
24	61.9	20.0	72	-27.7	32.4	121	41.5	349.5
25	-23.6	339.3	73	-22.9	351.7	126	-42.8	12.5
26	-48.7	344.5	74	10.3	32.5	127	-41.3	1.1
27	-55.0	347.2	75	6.6	350.0	128	-49.0	43.6
28	68.8	53.3	76	1.9	352.3	132	-11.8	68.9
29	-11.3	323.1	77	0.1	352.8	134	58.3	320.6
30	-14.8	321.3	78	24.7	358.0	135	73.2	337.9
31	-28.0	335.3	79	-30.1	13.2	136	59.2	335.3
32	-17.6	327.6	80	-37.3	60.2	137	-9.0	335.5
33	-25.9	322.2	81	-42.9	79.2	141	-31.6	328.1
34	-24.7	318.4	82	15.3	355.4	144	48.1	329.7
36	52.6	337.9	83	-6.4	52.0	146	-39.6	325.1
37	52.5	328.8	84	-13.9	50.6	147	-40.5	319.9
38	52.2	321.0	85	-18.9	68.9	148	-36.4	315.8
39	47.8	314.2	86	10.8	66.2	149	-17.6	343.6
40	42.4	334.2	88	2.8	53.2	152	-64.1	347.2
41	42.5	313.6	89	-7.1	39.2	153	-76.8	15.2
42	-11.2	341.5	90	-15.3	33.2	154	-77.3	8.4
43	15.7	323.9	92	-40.5	338.2	157	57.0	46.1
44	14.1	331.9	93	-43.1	337.2	163	-5.4	351.0
45	5.1	331.4	94	-46.3	339.1	164	-5.8	359.0
46	-1.7	331.2	95	49.4	339.7	165	1.0	340.5
47	25.6	318.2	96	48.3	339.4	166	-19.8	14.2
48	28.6	329.5	97	49.8	337.7	167	-21.6	12.1
49	26.0	342.1	98	68.8	347.0	168	-11.0	11.9

GANYMEDE Table 15--Continued

Point	Lat.	Long.	Point	Lat.	Long.	Point	Lat.	Long.
169	-10.0	6.0	244	17.1	138.7	320	-4.5	16.3
170	-17.5	7.2	245	38.4	105.0	321	-0.9	11.2
171	-17.8	1.7	248	20.6	48.9	322	1.8	6.7
175	-12.8	337.5	260	58.8	37.5	323	0.5	359.9
178	19.2	15.1	271	51.6	10.1	324	12.5	23.2
179	17.1	22.0	277	51.9	33.4	325	-7.5	331.1
180	13.3	16.4	278	53.3	26.6	326	-29.2	22.9
181	8.3	11.0	279	55.2	21.9	327	-26.9	8.3
182	12.6	25.8	280	40.7	36.2	328	-25.6	359.7
183	28.8	29.2	281	7.5	135.6	329	37.8	48.0
202	-15.0	339.4	282	4.1	156.2	330	33.5	44.7
203	-21.2	345.4	283	2.5	139.3	337	-49.7	7.7
205	13.3	344.6	284	33.1	145.0	338	-55.4	7.9
206	16.9	340.9	285	40.7	168.9	339	-62.6	32.1
207	-0.1	337.5	286	33.7	167.1	347	21.1	68.0
208	8.3	345.4	287	30.5	166.5	348	17.1	50.2
210	-1.6	341.0	288	20.5	164.8	349	-3.4	46.2
212	-17.9	338.3	289	17.3	189.2	350	-2.6	35.2
218	32.4	235.8	290	23.6	198.9	351	0.1	32.2
219	37.8	197.4	291	20.9	194.8	352	14.4	0.8
220	34.8	150.7	299	26.0	218.9	353	9.3	357.9
221	-3.9	185.5	302	7.2	232.3	354	5.2	356.5
223	6.7	163.2	304	46.2	214.5	355	1.7	355.0
224	3.1	162.6	305	7.9	168.4	356	40.2	22.7
225	-6.3	192.7	306	5.8	184.0	357	22.8	332.4
226	11.7	189.4	309	-8.6	48.4	358	21.7	341.0
227	3.2	191.2	310	-9.7	39.4	359	-13.2	335.6
228	24.0	176.8	311	-14.1	39.6	361	17.7	341.4
229	23.2	170.1	312	-20.7	40.9	362	19.4	341.0
230	32.2	171.6	313	-23.2	40.7	363	19.2	345.6
231	9.4	215.2	314	-25.9	43.3	364	17.5	351.9
232	10.3	228.6	315	-29.9	48.2	366	49.4	333.6
233	4.8	202.0	316	-36.2	71.4	367	47.6	334.5
241	17.0	109.5	317	-1.7	41.7	369	3.1	329.8
242	12.5	99.0	318	-5.7	30.4	370	0.2	341.6
243	10.0	128.7	319	-6.0	25.5			

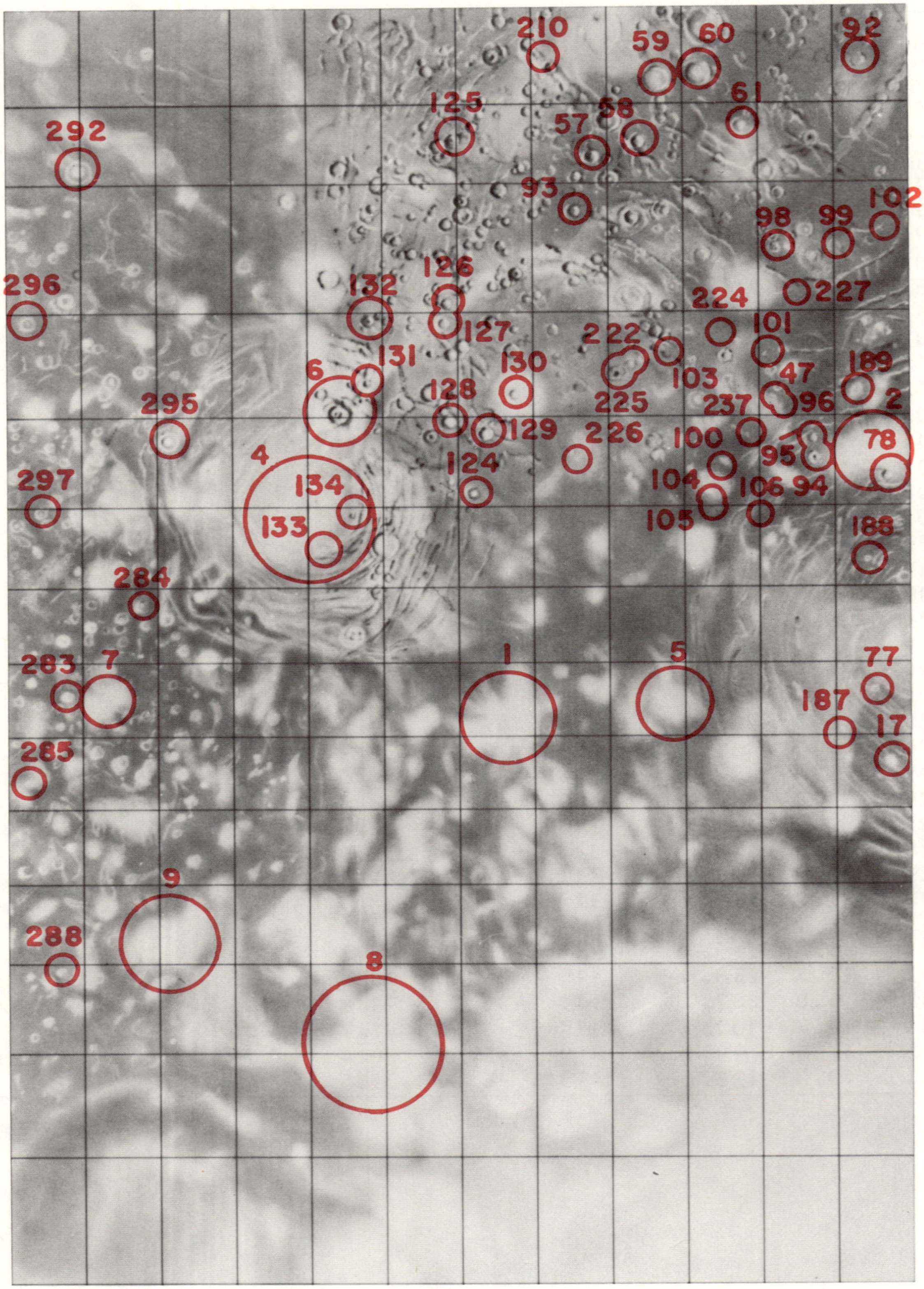

Fig. 29 — Callisto: Mercator map with control points identified in the region west of the prime meridian

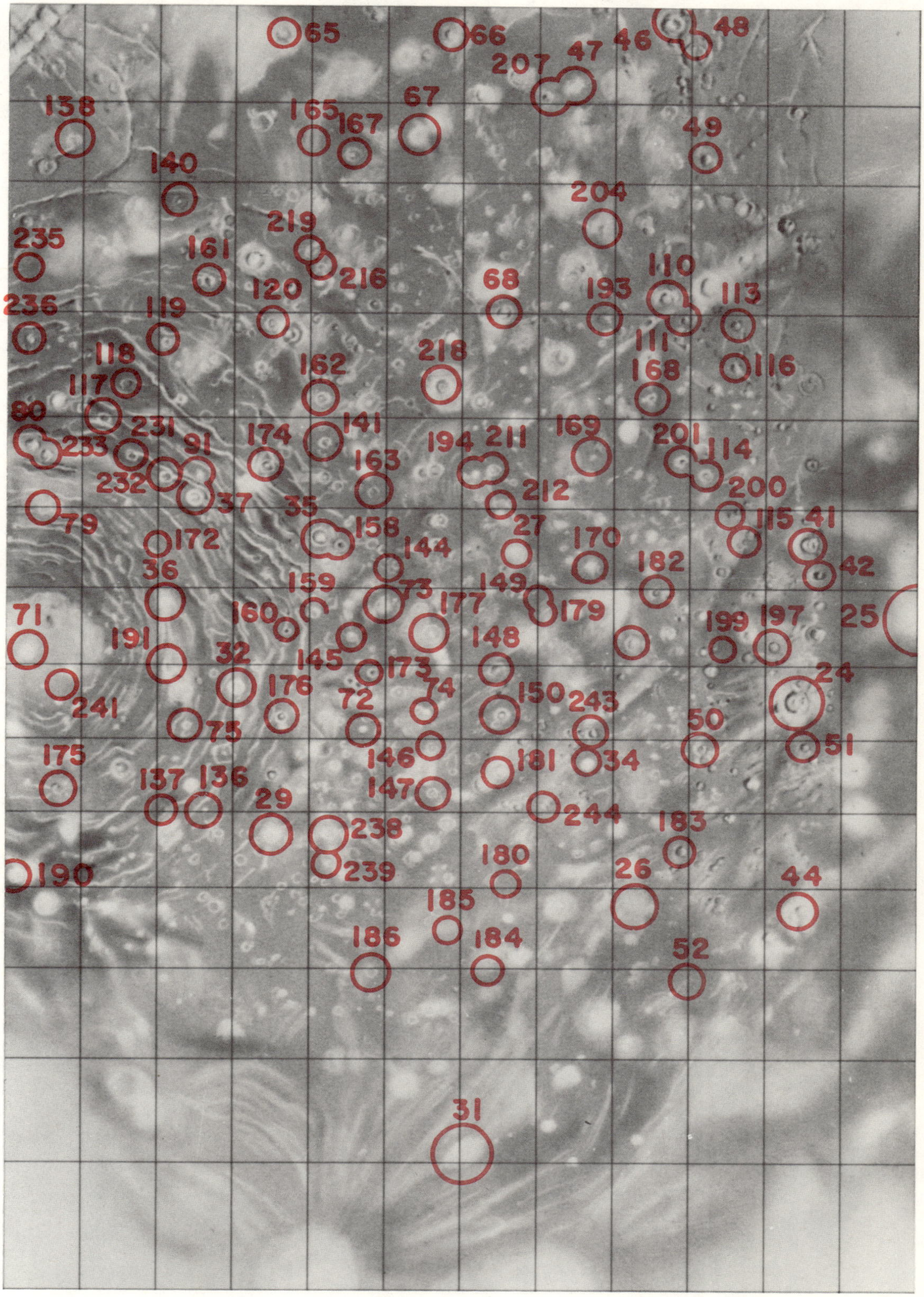

Fig. 30 — Callisto: Mercator map with control points identified in the region of the prime meridian

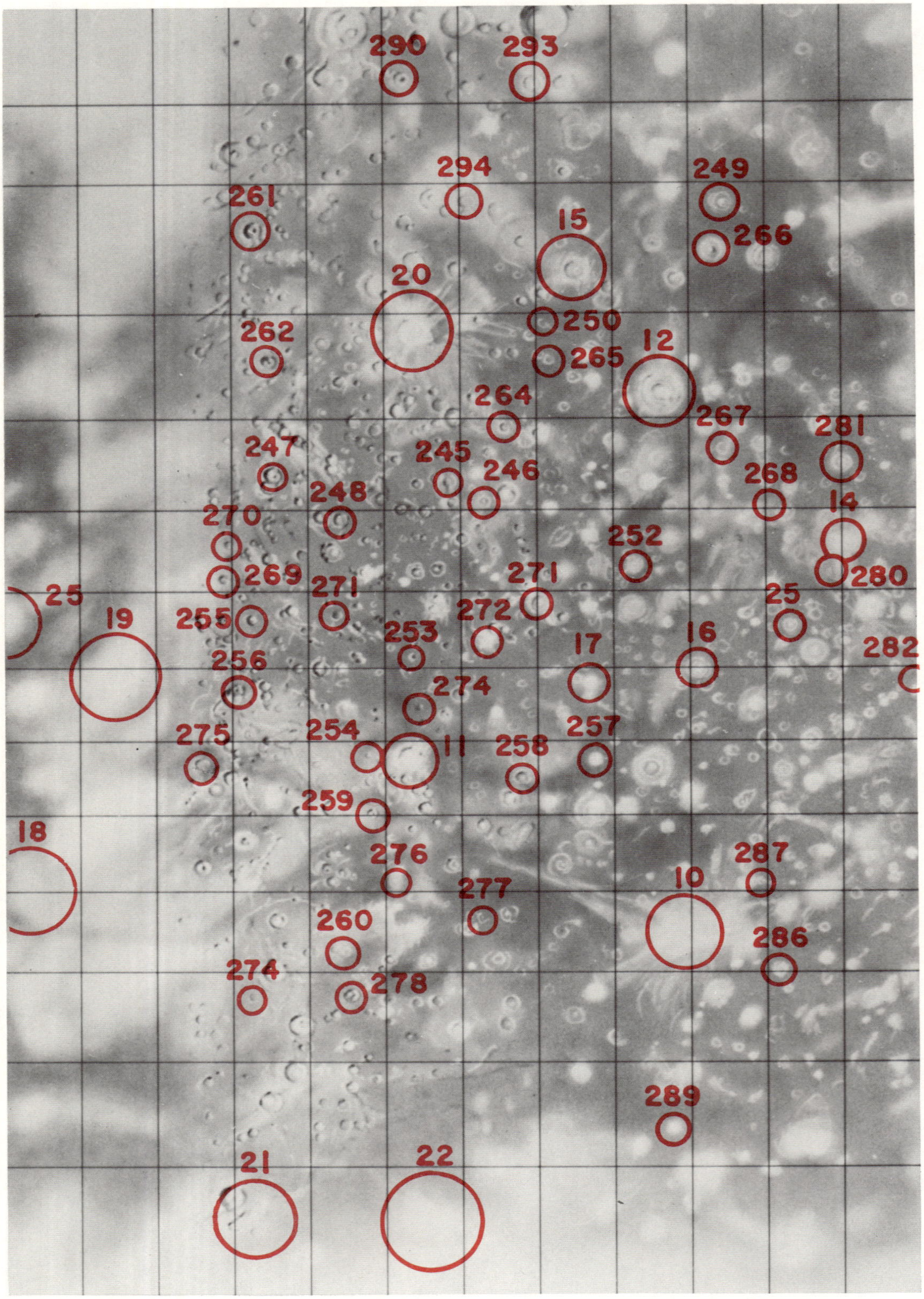

Fig. 31 — Callisto: Mercator map with control points identified in the region east of the prime meridian

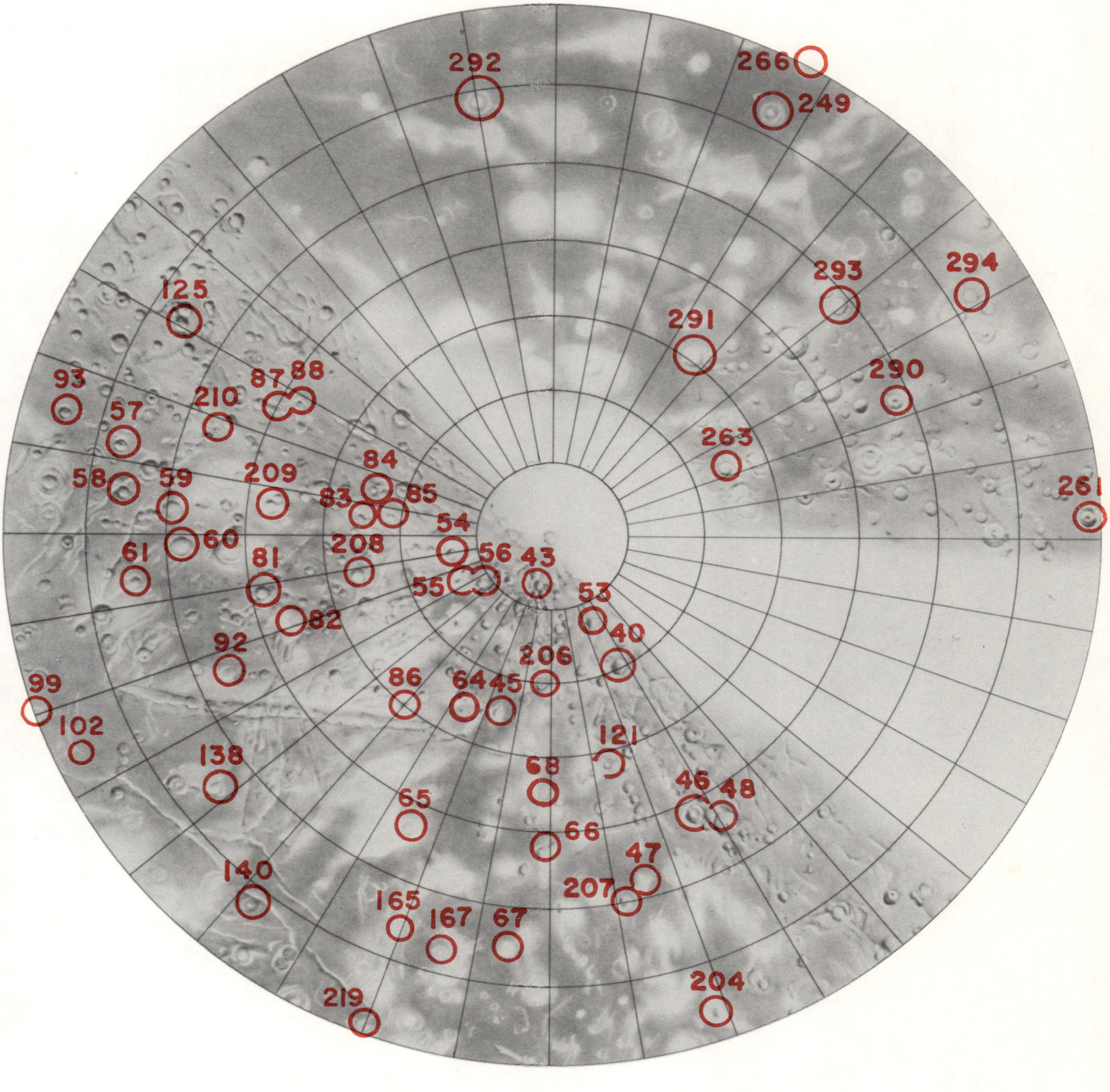

Fig. 32 — Callisto: Stereographic map with control points identified in the region of the north pole

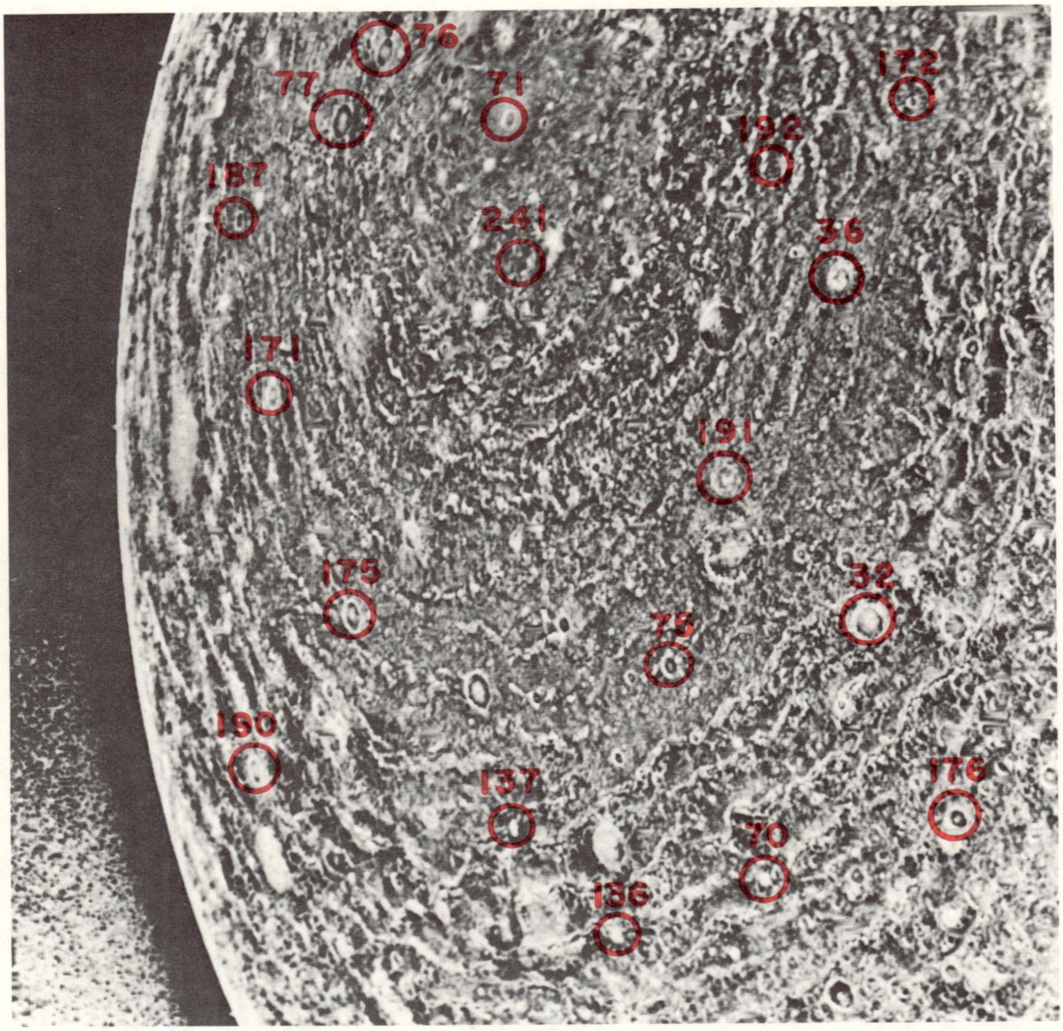

Fig. 33 — Callisto: Near-encounter picture with control points identified

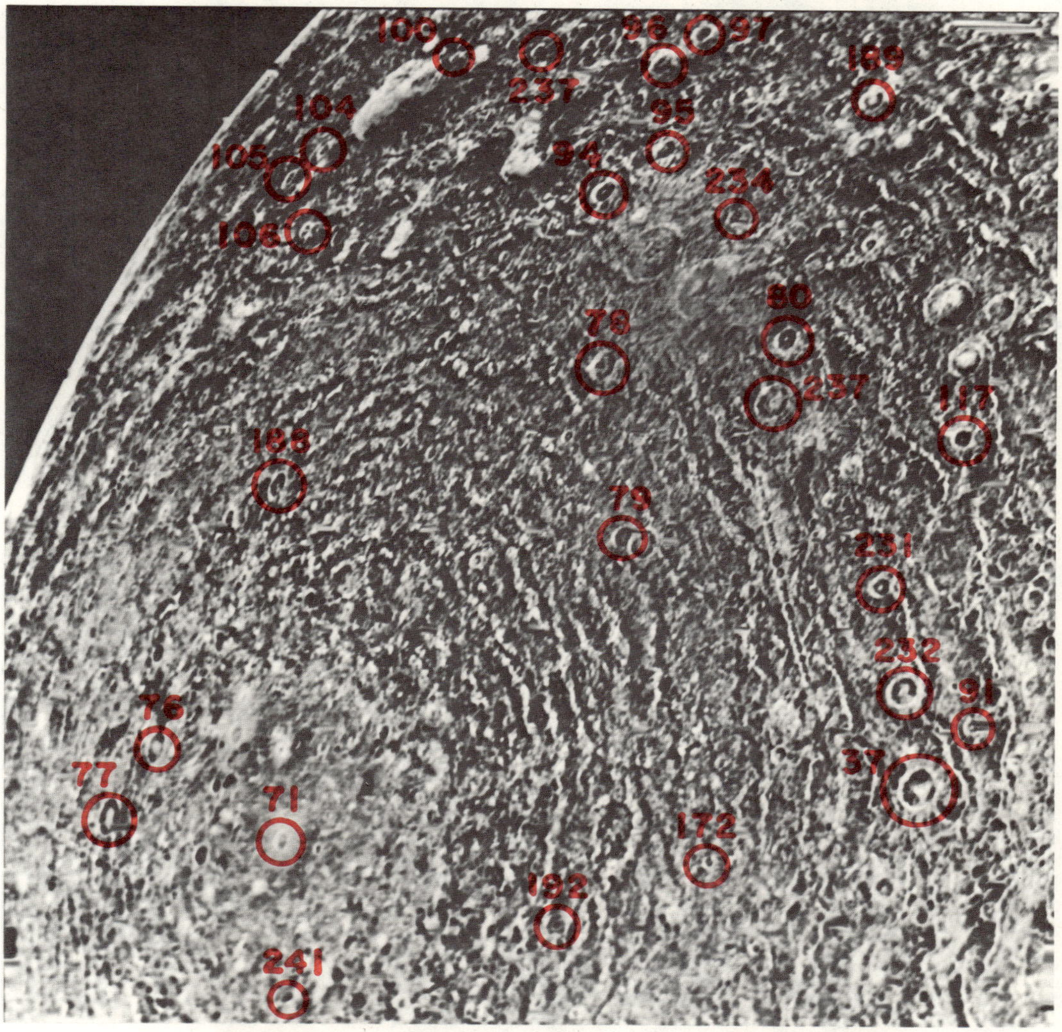

Fig. 34 — Callisto: Near-encounter picture with control points identified

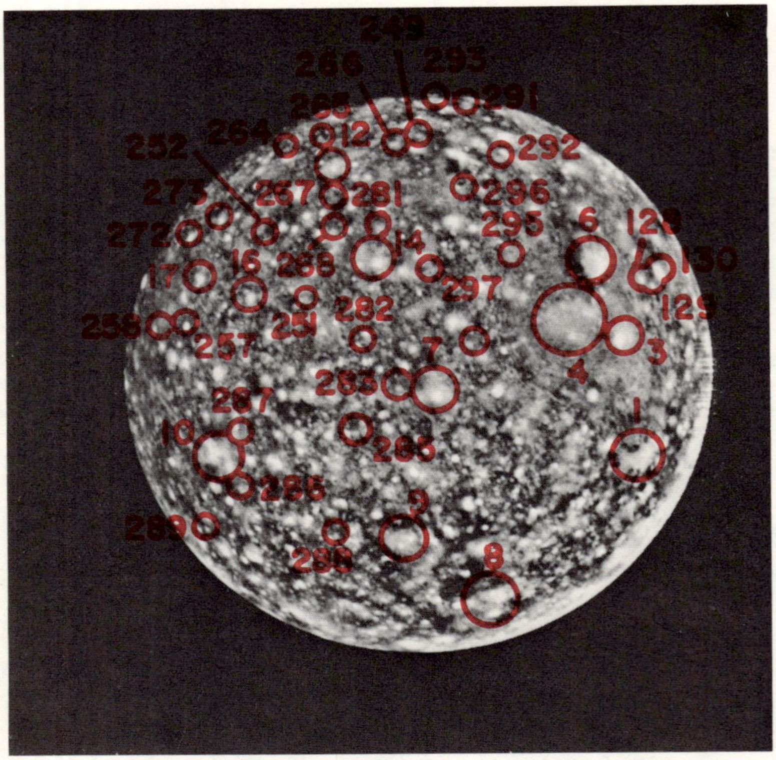

Fig. 35 — Callisto: Limb picture with control points identified

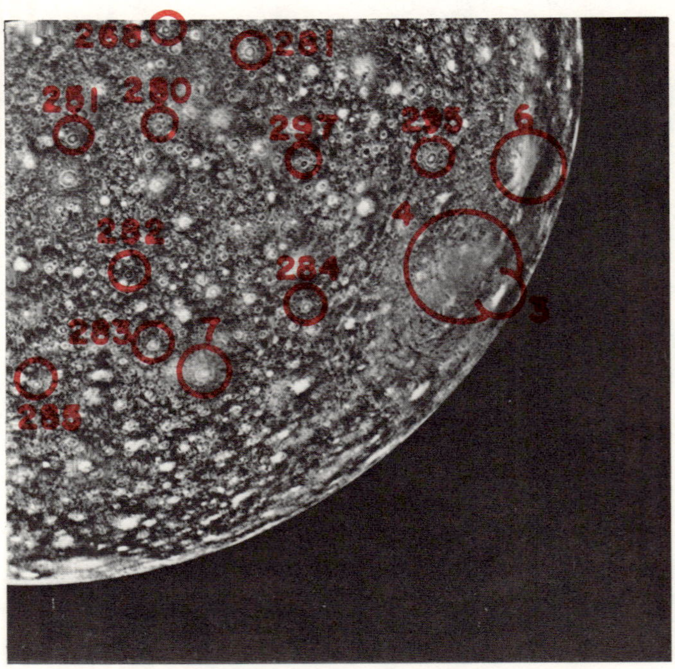

Fig. 36 — Callisto: Far-encounter picture with control points identified

Table 16

CALLISTO: COORDINATES OF CONTROL POINTS

Point	Lat.	Long.	Point	Lat.	Long.	Point	Lat.	Long.
1	-3.3	96.3	44	-21.9	316.3	83	77.2	93.9
2	30.9	39.3	45	76.0	3.0	84	77.1	107.9
3	18.4	109.2	46	67.2	322.5	85	79.4	103.1
4	19.3	123.3	47	64.7	336.8	86	72.9	46.6
5	-0.3	71.2	48	65.6	321.6	87	69.6	121.5
6	33.7	115.1	49	60.1	320.6	88	70.5	125.2
7	-4.0	144.5	50	-2.1	329.1	89	69.7	126.8
8	-48.1	101.2	51	-20.7	237.4	91	37.9	35.5
9	-39.9	135.6	52	-32.3	331.2	92	68.4	68.4
10	-31.3	196.9	52	-32.4	331.1	93	58.2	105.4
11	-1.0	238.8	53	79.2	321.3	94	38.1	70.2
12	33.4	198.6	54	83.4	81.0	95	40.4	70.4
14	19.3	175.3	55	82.7	52.9	96	40.9	76.0
15	45.9	221.0	56	84.3	39.4	97	42.5	76.8
16	4.4	197.8	57	63.2	108.6	98	59.7	80.9
17	3.3	213.8	58	64.3	102.5	100	32.8	84.8
18	-12.5	282.7	59	67.2	97.0	101	48.2	80.0
19	8.9	273.1	60	67.9	91.0	103	47.7	95.5
20	40.9	241.4	61	65.3	85.2	104	27.1	84.9
21	-47.3	267.7	62	72.0	351.3	105	25.7	84.9
22	-50.9	246.6	63	74.6	24.3	106	26.3	80.0
23	-21.1	315.5	64	76.0	19.1	108	-23.5	49.2
24	7.0	314.3	65	68.9	17.7	109	-25.0	28.5
25	17.7	290.6	66	68.3	347.8	110	51.2	328.0
26	-20.2	338.7	67	64.0	357.5	111	49.3	324.6
27	25.3	350.1	68	52.5	348.8	112	44.8	329.9
28	44.2	358.3	69	58.1	337.9	113	48.8	319.6
29	-8.0	26.9	70	1.6	31.3	114	34.9	324.7
30	44.7	38.8	71	17.7	56.1	115	26.5	321.2
31	-47.6	0.8	72	5.2	14.4	116	45.6	320.5
32	12.6	31.2	73	21.0	10.1	117	44.4	46.0
34	-2.0	342.7	74	7.1	5.4	118	48.3	43.5
35	30.0	18.6	75	6.6	37.9	119	50.8	37.3
36	22.2	38.8	76	15.0	64.2	120	50.9	21.9
37	35.3	35.6	77	11.9	64.2	121	70.7	335.9
38	67.6	359.2	78	36.0	61.6	122	48.9	87.8
39	63.2	329.1	79	33.7	53.8	123	28.6	110.1
40	75.4	321.6	80	41.7	56.4	124	36.8	118.5
41	25.3	314.4	81	71.6	82.6	125	63.2	124.3
42	21.7	313.5	82	72.4	73.7	126	54.2	121.3

CALLISTO Table 16--Continued

Point	Lat.	Long.	Point	Lat.	Long.	Point	Lat.	Lat.
127	52.3	120.9	174	39.1	24.8	214	48.6	7.1
128	42.8	118.9	175	-1.2	52.6	215	60.1	22.7
129	41.8	114.3	176	7.8	25.1	216	55.8	15.6
130	44.7	109.9	177	16.6	4.0	217	53.2	12.3
131	46.5	133.6	178	14.2	336.2	218	47.8	1.9
132	52.4	132.8	179	17.0	348.1	219	56.3	19.9
133	29.3	135.9	180	-17.3	353.9	220	63.4	37.5
134	31.8	133.6	181	-1.4	355.4	221	58.6	25.3
135	37.6	134.5	182	20.3	332.6	222	46.3	100.6
136	-4.2	36.4	183	-17.6	331.6	223	53.3	78.0
137	-3.2	42.0	184	-29.4	357.1	224	50.2	88.5
138	64.9	48.9	185	-22.3	1.7	225	42.6	103.2
139	62.9	47.8	186	-27.4	10.0	226	32.7	111.6
140	61.2	34.7	187	5.2	67.5	227	7.0	3.0
141	40.4	17.2	188	23.5	68.8	228	23.5	19.9
142	43.4	3.7	189	47.5	66.7	229	28.5	23.4
143	34.6	1.4	190	-10.9	56.4	230	33.0	23.9
144	25.6	9.3	191	13.7	39.0	231	39.1	43.2
145	18.4	15.0	192	23.6	44.0	232	37.2	39.0
146	1.4	5.1	193	47.5	335.4	233	40.2	54.1
147	-3.3	5.2	194	35.6	357.3	234	41.9	64.3
148	11.8	355.4	195	45.7	315.2	235	57.0	56.3
149	21.0	348.6	196	11.9	359.0	236	52.2	55.8
150	5.1	354.0	197	12.9	318.1	237	36.3	81.6
158	29.1	15.9	198	25.4	325.1	238	-6.8	18.9
159	21.3	20.1	199	12.6	323.8	239	-12.3	19.5
160	22.0	22.7	200	29.4	322.4	240	4.2	35.9
161	56.0	31.3	201	36.9	326.7	241	14.3	51.4
162	44.5	16.6	202	28.2	326.8	242	3.5	333.2
163	34.8	10.5	203	52.1	330.2	243	1.8	342.5
164	50.2	1.3	204	57.4	333.9	244	-5.8	349.4
165	62.6	15.9	205	64.4	333.3	298	-6.0	39.0
166	80.1	58.7	206	76.4	352.0	299	-17.6	38.2
167	61.0	7.7	207	64.4	340.0	300	-0.7	37.3
168	43.0	330.6	208	76.4	82.1	4001	73.4	36.2
169	38.0	338.2	209	71.9	104.2	4002	64.8	68.0
170	25.2	340.5	210	66.7	114.7	4003	-53.9	21.0
171	2.3	60.9	211	36.6	353.9	4005	43.0	135.8
172	28.8	40.6	212	31.7	352.7	4006	67.9	137.1
173	12.0	10.8	213	44.7	359.7	4007	60.5	116.2

CALLISTO Table 16--Continued

Point	Lat.	Long.	Point	Lat.	Long.	Point	Lat.	Long.
4008	66.7	122.1	4031	56.5	24.7	4050	10.0	325.0
4009	47.7	103.4	4032	65.4	22.3	4051	22.5	323.4
4010	52.3	101.6	4034	50.2	19.5	4052	51.9	322.9
4012	62.1	105.7	4035	54.8	15.4	4053	57.3	317.9
4013	65.2	106.7	4036	57.4	11.5	4054	65.4	320.8
4015	70.0	91.3	4037	-10.5	4.8	4064	76.3	136.7
4017	63.7	81.6	4038	19.3	5.1	4065	74.5	111.7
4018	63.6	78.0	4039	30.0	11.3	4066	77.3	78.0
4019	65.2	76.4	4040	44.1	10.3	4067	72.3	64.4
4020	-13.7	54.6	4041	-2.5	358.1	4068	78.5	19.1
4022	46.6	52.6	4042	53.4	351.3	4069	80.4	346.3
4023	60.2	56.1	4043	64.9	347.8	4070	72.6	315.1
4024	64.1	52.7	4044	65.2	343.2	4071	55.5	314.0
4025	0.5	45.4	4045	-20.9	349.6	5076	15.8	64.3
4026	-35.4	43.3	4046	45.7	342.6	5106	26.7	80.5
4029	58.5	32.3	4047	56.0	328.4	5240	5.2	35.4
4030	37.0	28.6	4048	56.6	329.1	5241	15.2	51.4

BIBLIOGRAPHY

Benesh, M., and P. Jepsen (1978): *Voyager Imaging Science Subsystem Calibration Report*, Jet Propulsion Laboratory, No. 618-802, July 31, 1978.

Davies, M. E. (1972): "Coordinates of Features on the Mariner 6 and 7 Pictures of Mars," *Icarus*, Vol. 17, No. 1, August 1972, p. 116.

Davies, M. E., and D. W. G. Arthur (1973): "Martian Surface Coordinates," *J. Geophys. Res.*, Vol. 78, No. 20, July 10, 1973, p. 4355.

Davies, M. E., and R. M. Batson (1975): "Surface Coordinates and Cartography of Mercury," *J. Geophys. Res.*, Vol. 80, No. 17, June 10, 1975, p. 2417.

Lieske, J. H. (1979): "Poles of the Galilean Satellites," *Astron. Astrophys.*, Vol. 75, 1979, p. 158.

Peale, S. J. (1977): "Rotation Histories of the Natural Satellites," *Planetary Satellites*, Joseph A. Burns (ed.), University of Arizona Press, 1977.

Smith, B. A., et al. (1977): "Voyager Imaging Experiment," *Space Science Reviews*, Vol. 21, No. 2, November 1977, p. 103.

Smith, B. A., et al. (1979a): "The Jupiter System through the Eyes of Voyager 1," *Science*, Vol. 204, June 1, 1979, p. 951.

Smith, B. A., et al. (1979b): "The Galilean Satellites and Jupiter: Voyager 2 Imaging Science Results," *Science*, Vol. 206, November 23, 1979, p. 927.

Transactions of the International Astronomical Union (1979): Proceedings of the Seventeenth General Assembly, Vol. XVIIB, Montreal, 1979 (to be published).

RAND/R-2532-JPL/NA[illegible]